W0107500

Lecture Notes in Physics

New Series m: Monographs

Editorial Board

H. Araki, Kyoto, Japan
E. Brézin, Paris, France
J. Ehlers, Potsdam, Germany
U. Frisch, Nice, France
K. Hepp, Zürich, Switzerland
R. L. Jaffe, Cambridge, MA, USA
R. Kippenhahn, Göttingen, Germany
H. A. Weidenmüller, Heidelberg, Germany
J. Wess, München, Germany
J. Zittartz, Köln, Germany

Managing Editor

W. Beiglböck
Assisted by Mrs. Sabine Lehr
c/o Springer-Verlag, Physics Editorial Department II
Tiergartenstrasse 17, D-69121 Heidelberg, Germany

Springer-Verlag
Berlin Heidelberg GmbH

The Editorial Policy for Monographs

The series Lecture Notes in Physics reports new developments in physical research and teaching - quickly, informally, and at a high level. The type of material considered for publication in the New Series m includes monographs presenting original research or new angles in a classical field. The timeliness of a manuscript is more important than its form, which may be preliminary or tentative. Manuscripts should be reasonably self-contained. They will often present not only results of the author(s) but also related work by other people and will provide sufficient motivation, examples, and applications.
The manuscripts or a detailed description thereof should be submitted either to one of the series editors or to the managing editor. The proposal is then carefully refereed. A final decision concerning publication can often only be made on the basis of the complete manuscript, but otherwise the editors will try to make a preliminary decision as definite as they can on the basis of the available information.
Manuscripts should be no less than 100 and preferably no more than 400 pages in length. Final manuscripts should preferably be in English, or possibly in French or German. They should include a table of contents and an informative introduction accessible also to readers not particularly familiar with the topic treated. Authors are free to use the material in other publications. However, if extensive use is made elsewhere, the publisher should be informed. Authors receive jointly 50 complimentary copies of their book. They are entitled to purchase further copies of their book at a reduced rate. As a rule no reprints of individual contributions can be supplied. No royalty is paid on Lecture Notes in Physics volumes. Commitment to publish is made by letter of interest rather than by signing a formal contract. Springer-Verlag secures the copyright for each volume.

The Production Process

The books are hardbound, and quality paper appropriate to the needs of the author(s) is used. Publication time is about ten weeks. More than twenty years of experience guarantee authors the best possible service. To reach the goal of rapid publication at a low price the technique of photographic reproduction from a camera-ready manuscript was chosen. This process shifts the main responsibility for the technical quality considerably from the publisher to the author. We therefore urge all authors to observe very carefully our guidelines for the preparation of camera-ready manuscripts, which we will supply on request. This applies especially to the quality of figures and halftones submitted for publication. Figures should be submitted as originals or glossy prints, as very often Xerox copies are not suitable for reproduction. For the same reason, any writing within figures should not be smaller than 2.5 mm. It might be useful to look at some of the volumes already published or, especially if some atypical text is planned, to write to the Physics Editorial Department of Springer-Verlag direct. This avoids mistakes and time-consuming correspondence during the production period.
As a special service, we offer free of charge LaTeX and TeX macro packages to format the text according to Springer-Verlag's quality requirements. We strongly recommend authors to make use of this offer, as the result will be a book of considerably improved technical quality.
Manuscripts not meeting the technical standard of the series will have to be returned for improvement.
For further information please contact Springer-Verlag, Physics Editorial Department II, Tiergartenstrasse 17, D-69121 Heidelberg, Germany.

Alexander Bach

Indistinguishable Classical Particles

Springer

Author

Alexander Bach
Friedenstraße 12
D-48231 Warendorf
Germany

Die Deutsche Bibliothek - CIP-Einheitsaufnahme

Bach, Alexander:
Indistinguishable classical particles / Alexander Bach.
(Lecture notes in physics : N.s. M, Monographs ; 44)
ISBN 978-3-662-14165-6 ISBN 978-3-540-49624-3 (eBook)
DOI 10.1007/978-3-540-49624-3

NE: Lecture notes in physics / M

ISSN 0940-7677 (Lecture Notes in Physics. New Series m: Monographs)
ISBN 978-3-662-14165-6

This work is subject to copyright. All rights are reserved, whether the whole or part of the material is concerned, specifically the rights of translation, reprinting, re-use of illustrations, recitation, broadcasting, reproduction on microfilms or in any other way, and storage in data banks. Duplication of this publication or parts thereof is permitted only under the provisions of the German Copyright Law of September 9, 1965, in its current version, and permission for use must always be obtained from Springer-Verlag Berlin Heidelberg GmbH.
Violations are liable for prosecution under the German Copyright Law.

© Springer-Verlag Berlin Heidelberg 1997
Originally published by Springer-Verlag Berlin Heidelberg New York in 1997
Softcover reprint of the hardcover 1st edition 1997

The use of general descriptive names, registered names, trademarks, etc. in this publication does not imply, even in the absence of a specific statement, that such names are exempt from the relevant protective laws and regulations and therefore free for general use.

Typesetting: Camera-ready by author
Cover design: *design & production* GmbH, Heidelberg
SPIN: 10517847 55/3144-543210 - Printed on acid-free paper

Dedicated to the memory of

J. KAMPHUSMANN

Professor of Physics, Westfälische–Wilhelms–Universität, Münster.

Table of Contents

1. Introduction

Physicists realized the existence of indistinguishable particles at approximately the same time (1925-1926) when the foundations of modern quantum theory were established. Because of this coincidence the concept of indistinguishability immediately was brought into such a close connection with quantum mechanics that the foundations of this concept were exclusively confined to the new theory. While on the one hand the theoretical origins of indistinguishability in the early theory of blackbody radiation were forgotten, on the other hand the conceptual foundations of indistinguishability in the modern theory remained unsatisfactory and unclear. As a consequence, DELBRÜCK concluded that 'the loss of the category of identity should be considered as an additional characteristic of quantum mechanics, and perhaps its most eerie one' [48, p. 470].

Until today – in the familiar textbooks – the different but related concepts of identity and indistinguishability are logically not clearly separated, mixed with epistemological explanations such as the possibility to attach labels, to guarantee a redetection, to identify particles by paths, or ontological assertions concerning the non/existence of certain fundamental indistinguishable events or indiscernible objects. Moreover, if a quantum system is given that is invariant under a certain symmetry the consideration of any abelian subalgebra cannot break this symmetry (because this is no empirical process) – precisely, this however happens for indistinguishable particles because of an inappropriate terminology. Finally, I am not even aware of a clear definition: 'particles are called indistinguishable particles if .....'. According to JAMMER 'the subject, bearing upon physics and philosophy alike, seems to have not yet been fully explored from the epistemological and logical point of view' [94, p. 335–336].

To substantiate this claim and to fix my viewpoint I collect some obvious contradictions.

1.1 The Loss of Compatibility

1.1.1 Maxwell–Boltzmann Statistics

It seems to be an unquestionable fact, explained in any textbook on elementary statistical mechanics, that the particles of MB (Maxwell–Boltzmann) statistics are distinguishable. In the familiar statistical scheme where n identical particles are distributed onto d cells MB statistics is defined as follows [144, 62]. We fix a measurable space (Ω, F) and introduce configuration random variables $J_i : \Omega \rightarrow \{1, \ldots, d\}$. Here $[J_i = j], 1 \leq i \leq n, 1 \leq j \leq d$, denotes the event that particle i is in cell j. Different statistics are characterized by different probability measures P on (Ω, F). MB statistics is defined by equal *a priori* probabilities of the d^n configurations $\mathbf{j} \in \{1, \ldots, d\}^n$

$$P_{MB}(\mathbf{J} = \mathbf{j}) = (1/d)^n. \tag{1.1}$$

In technical terms, the random variables J_i are i.i.d. and J_1 is uniformly distributed.

Distinguishability or identifiability of the particles in this context obviously means that we are able to ascertain, for any particle, the cell in which it exists. According to Definition (1.1), however, all configurations are equally probable. In particular, P_{MB} is invariant under any permutation of the particles in the sense that

$$P_{MB}(J_i = j_i, 1 \leq i \leq n) = P_{MB}(J_i = j_{\pi(i)}, 1 \leq i \leq n) \tag{1.2}$$

holds for any permutation π of the integers $\{1, \ldots, n\}$. In technical terms, the random variables $J_1, \ldots J_n$ are interchangeable, or – equivalently – the induced measure on $\{1, \ldots, d\}^n$ is symmetric.

Therefore it is impossible to infer from the information available which particle is in which cell. Irrespective of any formalization of the concept of indistinguishability the particles of Maxwell–Boltzmann statistics are indistinguishable.

1.1.2 Indistinguishable Particles in 1868

Before the invention of quantum theory, BOSE [35] in 1924 established BE (Bose–Einstein) statistics for light quanta and EINSTEIN [58] in 1925 extended this 'ansatz' to account for massive bosons. In 1926, without using the methods of quantum theory, FERMI [64] generalized this approach, including the exclusion principle. These results, however, were immediately incorporated in the form of symmetric and antisymmetric wave functions by HEISENBERG [88] and DIRAC [53] and eventually systematically by WIGNER [154] into the new quantum theory, such that their conceptual independence was not realized.

The logical independence of the concept of indistinguishable particles from the framework of quantum theory becomes more evident if we analyze the historical origin of BE statistics. BE statistics was introduced by BOLTZMANN in the period 1868–1877 as a discrete scheme to derive the exponential distribution (Boltzmann distribution) and to establish the entropy–probability relationship (cf. Chapter 5.1).

1.1.3 Quantum/Classical Conceptual Incompatibility

Let us, for the moment, adopt the traditional quantum viewpoint – identical particles are indistinguishable if they are characterized by symmetric observables – and compare it with the classical setting of BE and MB statistics with special attention to the concept of in/distinguishability.

For MB statistics we use the description given above. Moreover we introduce occupation number random variables $K_i : \Omega \to \{0, \ldots, n\}, i \leq i \leq d$. Here $[K_i = k]$ denotes the event that there are exactly k particles in cell i such that $\sum K_i = n$ holds P-a.e..

If configuration random variables $\mathbf{J}$ are already defined, the occupation number random variables are derived observables that are defined by

$$K_i = \sum_{m=1}^{n} 1_{[J_m = i]}. \tag{1.3}$$

E.g. for MB statistics we have

$$P_{MB}(\mathbf{K} = \mathbf{k}) = \binom{n}{k_1 \ldots k_d} (\frac{1}{d})^n. \tag{1.4}$$

BE statistics usually is defined by the uniform distribution of the occupation numbers $\mathbf{k} \in \{0, \ldots, n\}^d$ subject to the constraint $\sum k_i = n$

$$P_{BE}(\mathbf{K} = \mathbf{k}) = \binom{d+n-1}{n}^{-1}. \tag{1.5}$$

The traditional argument for the distinguishability of the particles of MB statistics and the indistinguishability of the particles for BE statistics (due to NATANSON (cf. Chapter 5.2)) is the following one.

- The particles of MB statistics are distinguishable since there exist configuration random variables $\mathbf{J}$, i.e. a description involving the individual particles.
- The particles of BE statistics are indistinguishable since a description involving the individual particles does not exist but only a description on a higher collective level by means of the occupations number random variables $\mathbf{K}$.

First of all, and most important in our context, this concept of indistinguishability (which I call the 'combinatorial concept of indistinguishability') has no logical connection with the quantum concept which is based upon the symmetry of observables. Moreover, this concept implies an ontological assertion concerning the nonexistence of certain events for BE statistics. Since no configuration observables exist the correlations of the particles of BE statistics cannot be evaluated (do not exist ?).

Two remarks are in order here. First, as mentioned before, the existence of the configuration random variables for MB statistics does not imply that we can distinguish these particles. Second, using the converse of the strategy that connects the occupation numbers with the configurations and imposing symmetry, we may postulate the permutation invariant distribution

$$P_{BE}(\mathbf{J}=\mathbf{j}) = \binom{n}{\kappa_1(\mathbf{j}) \ \dots \ \kappa_d(\mathbf{j})}^{-1} \binom{d+n-1}{n}^{-1}, \tag{1.6}$$

where for $1 \leq i \leq d$

$$\kappa_i(\mathbf{j}) = \sum_{m=1}^{n} \delta_{j_m,i}, \tag{1.7}$$

a formula which can be derived from the quantum theory of bosons (cf. Chapter 2.3).

A final remark on FD (Fermi–Dirac) statistics. Traditionally the set of occupation numbers $\mathbf{k}$ is confined to those events where any cell contains at most one particle. Instead of postulating the ontological nonexistence of multiple occupations, we prefer to assign probability zero to them and set, $d \geq n$,

$$P_{FD}(\mathbf{K}=\mathbf{k}) = \begin{cases} \binom{d}{n}^{-1}, & \text{if } \ \mathbf{k} \in \{0,1\}^d, \\ 0 & \text{else}, \end{cases} \tag{1.8}$$

a formula which may be derived from the quantum theory of fermions (cf. Chapter 2.3).

1.1.4 The Principle of Indistinguishability of Identical Particles

There is a familiar tradition to argue that identical particles in quantum theory are, in contrast to the situation in classical mechanics, in principle indistinguishable.

One line of arguments (see e.g. [127, 106]) makes use of the fact that in quantum theory there are no paths that allow us to identify a particle by its trajectory. Related, and well-known, is the assertion that 'if the wave functions of the [...] particles overlap to any extent, they intertwine so thoroughly that their identity is lost' [95, p. 276].

These arguments are useless. For, assume that some particles are (i) identical and (ii) that their wave functions overlap: nothing does imply that the wave function is anti/symmetric. Moreover, arguments based upon the temporal evolution are misleading since for the description of indistinguishable objects a time dependent setting is not necessary. Finally, in classical statistical mechanics we may have for two identical particles a symmetric probability distribution $p(t, \mathbf{x}_1, \mathbf{x}_2)$ in configuration space. In this situation the particles cannot be distinguished by their trajectories. I remark in conclusion that the principle of indistinguishability of identical particles violates the cluster law, that is, the existence of distinguishable identical particles.

1.2 Symmetry of the State

Obviously, indistinguishability is related to permutation invariance. The problem is to choose the appropriate reference object for this symmetry.

1.2.1 Symmetries

Symmetries and the connected conserved quantities (if they exist) are a fundamental tool in theoretical physics. Whereas the symmetries connected with Galilei invariance, reflection invariance or time reversal are discussed in classical mechanics (cf. e.g. [141]), invariance under permutations is not mentioned. In the C^*–algebraic formulation of quantum theory the invariance of a state under a general group of transformations is studied (cf. e.g. [61]) but permutation invariance is not considered explicitly. With two exceptions – BOGOLYUBOV as cited in [82] and BLOCHINZEW [32] – nowhere, as far as I am aware, the concept of the symmetry of the state is used to define indistinguishability. The analysis of this invariance the the goal of this work. After the preparation of the first draft of this work [13] symmetry of the state has been analyzed in quantum field theory (cf. e.g. [84]).

1.2.2 Symmetry of Observables and/or States

Disregarding those works which begin the introduction of indistinguishable particles by means of the discussion of scattering experiments, indistinguishability in quantum theory usually is defined by the symmetry of the observables (see e.g. [95, 117]).

The theory of irreducible representations of the symmetric group then implies that only an invariant subspace of the multiparticle Hilbert space is relevant (cf. e.g. [95]). Consequently, the theory is restricted to that subspace and, therefore, only statistical operators defined on this subspace – which are certainly symmetric statistical operators – are considered.

Accordingly, in the traditional treatment, for the description of indistinguishable particles symmetric operators for both the observables and the states are used. It is obvious, however, and has been shown explicitly, e.g. by

DRESDEN [54], that a definition based upon symmetric observables implies absolutely nothing concerning the state. Therefore the traditional procedure is sufficient but not necessary and implies a convenient but redundant description.

This redundancy was observed and explicitly avoided by HARTLE and TAYLOR [86] insofar as they used for the classification of indistinguishable particles into parastatistics symmetric observables but state vectors that are not symmetric.

In this work I concentrate on the other alternative of the traditional redundant description, namely on the symmetry of the state. This notion is general enough to embrace a concept of classical and quantum indistinguishable particles such that in abelian subalgebras the symmetry is preserved and transferred from the quantum to the classical domain.

Applying, in this setting, again the representation theory of the symmetric group, we can restrict the observables to symmetric observables. This explains why our results do not differ from the familiar ones. What is changed is the aspect, the object that, by definition, induces the symmetry. This offers a rich theory, the fruitful concept of extendibility and, in particular, de Finetti's theorem as a fundamental theoretical tool.

Before I proceed, I want to elucidate a fundamental misunderstanding. The unitary operators U_π, introduced in Section 2.1, are called permutation operators. This does not entail that these operators describe the process (in space and time) of permuting particles. As any element of an unitary representation of a discrete group such an operator has a passive meaning only (alias transformation), that is, the application of a permutation operator gives a description by means of new (permuted) names.

Moreover, it seems useful to resolve from the beginning a familiar misunderstanding concerning the seemingly contradiction between a concept that denies the possibility of an identification and a formalism which uses names (indices). Indistinguishability is an invariance principle that applies to the state of distinguished species of identical particles. To formulate this invariance in a general context – i.e. a context not restricted to this invariance – we first introduce indices for the particles and in a second step, – postulating invariance under permutations – we assert that these indices are irrelevant. In this context I emphasize that for indistinguishable particles it is misleading to call $P(J_1 = j_1, \ldots, J_n = j_n)$ the probability that particle i is in cell $j_i, 1 \leq i \leq n$. More precisely, it is the probability that one particle is in cell j_1, another one in cell j_2, ..., and another particle in cell j_n. Finally, I stress that the indices are no properties of the particles, they are a property of the description.

Compared to other symmetries (time reversal, space reflection, rotation invariance), invariance under permutations – as far as the observables are concerned – seems to go far beyond what is conceivable. Whereas 'the loss of identity' is well accepted as a phenomenon, nobody would interpret rotation invariance in terms of 'the loss of direction'.

To be explicit, I am concerned here with a discrete, passive symmetry concerning the state (statistical operators, probability measures). In the quantum part I confine myself to an analysis of this symmetry in an abstract Hilbert space setting and in the classical part I confine myself to discrete probability distributions and their limit laws.

Whereas in physics permutation symmetry of the state – up to recent exceptions in quantum field theory [84] – has not been analyzed explicitly, in classical probability there exists a well established theory of this symmetry, the theory of interchangeable random variables. Indistinguishable classical particles are characterized by a symmetric probability measure. From the physical and mathematical viewpoint it is more convenient to deal directly with the observables (random variables) of indistinguishable particles. These random variables, introduced by DE FINETTI in the late 20th, are called *interchangeable* (exchangeable) random variables. They generalize the familiar i.i.d. random variables. Unfortunately, owing to the misleading terminology of early quantum theory (distinguishable MB statistics) used by physicists, the identity of the the concept of interchangeable random variables to the concept of indistinguishable particles has not been recognized. Due to this situation 'for almost half a century, exchangeability has been considered an isolated branch of probability comprising essentially only one theorem, and interesting only because this sole theorem happens to be a pearl' [98].

1.2.3 Indistinguishability

Indistinguishable Classical Particles Have No Trajectories. The unconventional role of indistinguishable classical particles is best expressed by the fact that in a deterministic setting no indistinguishable particles exist, or – equivalently – that indistinguishable classical particles have no trajectories. Before I give a formal proof I argue as follows. Suppose they have trajectories, then the particles can be identified by them and are, therefore, not indistinguishable.

For our purposes it is sufficient to prove this statement in a time independent setting and to show that two indistinguishable classical particles (in one dimension) have no coordinates. We say that two particles have coordinates if their state is an element of the set

$$K = \mathrm{ex}\,(\,M_+^1(\mathbf{R}^2 \setminus D)\,), \tag{1.9}$$

where, to account for impenetrability, the 'diagonal'

$$D = \{(x,x) \in \mathbf{R}^2; x \in \mathbf{R})\} \tag{1.10}$$

is eliminated. Two indistinguishable particles are described by a symmetric probability measure on $\mathbf{R}^2$. The set of symmetric probability measures on

$\mathbf{R}^2$, denoted by $M^1_{+,\mathrm{sym}}(\mathbf{R}^2)$, is a simplex and the extreme points are given by the measures (cf. e.g. [27])

$$\begin{aligned} &\mathrm{ex}\,\{M^1_{+,\mathrm{sym}}(\mathbf{R}^2)\} \qquad (1.11)\\ &\quad = \{\mu_{x,y} \in M^1_+(\mathbf{R}^2)\,;\, \mu_{x,y} = \frac{1}{2}(\delta_{(x,y)} + \delta_{(y,x)})\,,(x,y) \in \mathbf{R}^2\}. \end{aligned}$$

From this set we have to eliminate again the diagonal contributions and obtain the set

$$S = \{\mu_{x,y} \in M^1_+(\mathbf{R}^2)\,;\, \mu_{x,y} = \frac{1}{2}(\delta_{(x,y)} + \delta_{(y,x)})\,,(x,y) \in \mathbf{R}^2\setminus D\}. \qquad (1.12)$$

Indistinguishable particles having coordinates are characterized by probability measures that belong to the set

$$S \cap K = \emptyset. \qquad (1.13)$$

Accordingly, classical indistinguishable particles have no coordinates and, in a time-dependent framework, no trajectories. Consequences of this phenomenon are discussed in [9].

Definition. If we define

Indistinguishability = Identity of the Particles + Symmetry of the State

then all contradictions mentioned above vanish or become meaningless.

Indistinguishability is a genuine probabilistic concept: Indistinguishability is not an intrinsic property of particles but a property of their state. Furthermore, it refers not to the indistinguishability of events (see for example [37]), but it means the indistinguishability (identity) of the probabilities of different (permuted) events. This presupposes that it is possible, in principle, to distinguish the events which cannot be distinguished by their probabilities. Identification of the events, however, requires to break the symmetry. Whereas for a system of few particles this is conceivable, for macroscopic systems this requires the identification of some 10^{23} particles. For macroscopic systems I consider, therefore, indistinguishability as an intrinsic property of the particles.

I briefly mention the advantages and disadvantages of this approach.

Disadvantages.

1. We loose – in the framework of elementary non-relativistic quantum theory – the category of indistinguishable particles. There exist only identical particles in a state where they are indistinguishable. *Indistinguishability is a property of the state, not of the particles.*
2. We loose – in the framework of elementary statistics – the class of distinguishable MB particles. Nonetheless, MB particles preserve their distinguished role since they constitute the class of *independent* indistinguishable particles.
3. We loose a quantum myth (a property unexplainable in a classical setting).

Advantages.

1. We obtain an unifying framework for indistinguishable particles in classical and quantum theory.
2. In this setting the particles of MB statistics are indistinguishable.
3. We can evaluate correlations for the particles of BE and FD statistics.
4. We obtain de Finetti's theorem as a fundamental theoretical tool.

1.3 Technical Remarks

The material presented here has its origin in an analysis of the emergence of the representing measure of classical states of the quantum harmonic oscillator (see [21]). It contains the fundamentals of the classical part of these investigations, namely the analysis of de Finetti's theorem in abelian subalgebras and tries to embed these special topics into a broader context. This explains both the selection of the themes and their presentation.

Since physicists are acquainted with indistinguishable quantum particles but not with indistinguishable classical particles, I have included a chapter on indistinguishable quantum particles, founded onto the symmetry of the state, where the fundamentals are presented. In that introducing chapter the fundamental probability distributions for indistinguishable classical particles are derived in abelian subalgebras. This chapter is independent from the rest of this work.

As far as the theory of indistinguishable classical particles is concerned, I confine myself to a discrete setting, i.e. discrete–valued interchangeable random variables and their various macroscopic and continuum limits. The treatment is elementary insofar as I confine myself to the statistical scheme where structureless identical particles are distributed onto cells. Moreover, besides de Finetti's classical theorem and a multivariate variant, I consider the Poisson limit and the central limit of de Finetti's theorem.

From the *physical viewpoint* this work is intended to give a conceptual analysis of the notion of indistinguishability and to provide the reader with the basic tools for the description of indistinguishable classical particles. The collection of results should be considered as a preparation for the quantum theory of the classical states of the harmonic oscillator. As far as this generalization is concerned, I refer the reader to the quantum de Finetti theorem [93], the derivation of the integral representations of classical states [23, 24] and the quantum probabilistic discretization of the harmonic oscillator [21].

From the *mathematical viewpoint* this work is an elementary introduction to the theory of interchangeable random variables. Its emphasis is on the methodological side, collecting results that are scattered in the literature from an unifying viewpoint. This introduction to the physical applications of interchangeable random variables includes a rich class of elementary probability distributions (multivariate Pólya-Brillouin distributions, polyhypergeometric distributions, Dirichlet distributions, negative binomial distributions, gamma distributions) and basic examples for Markov processes, non Markovian processes, and martingales. It seems that an introduction to elementary probability theory is possible relying entirely on interchangeable random variables.

I remark that by the LLN for a sequence of random variables $X_i, i \in \mathbf{N}$, I always understand the convergence in distribution of the sequence $\sum^n X_i/n$.

A final remark on *proofs*. These can be divided into two classes. Technical proofs are cited from the literature and presented only if I found no reference. On the other hand, proofs which I consider as necessary for the development of the concepts under consideration are included. Moreover, *problems* are not exercises but open questions.

The contents of this work are obvious from the table of contents and are not repeated here.

This is a revised version of an earlier manuscript [13] written in 1989. I have completely rewritten the quantum part whereas the classical part has been both enlarged (as far as the results are concerned) and condensed (as far as details are concerned).

Acknowledgements. It is a pleasure to thank the members of the Seminar 'Statistische Physik' (Münster 1983-1986) and its organizers, the late Professor J. KAMPHUSMANN and Professor D. PLACHKY, for their interest and their help in the development of an unconventional idea in an intransigent and passive environment where the fact that MB statistics is described by a symmetric probability measure and its implications never have been understood. The efficiency of this Seminar, however, resulted from the collaboration of physicists and mathematicians.

In particular, I thank H. FRANCKE for his cooperation over many years. H. BLANCK analyzed the multivariate de Finetti theorem and its quantum generalization [31]. Many discussions on the conceptual foundations of physical theories with H. PIEHLER have been an invaluable help. Moreover, I thank U. LÜXMANN-ELLINGHAUS and A. SRIVASTAV for their cooperation concerning the quantum generalizations [23, 24]. Finally, I thank the staff of the 'ODEON' where I developed some ten years ago the foundations for a new theory of indistinguishability.

This project started in 1984 with the idea of a quantum generalization of the theory presented here [8]. I thank Professor R.L. HUDSON for drawing my attention in 1981 to the integral representations of the quantum de Finetti theorem [93]. Some sections of this work have been written while I was visiting several times in the period 1985–1988 the Centro Matematico V. Volterra of the Università di Roma II, Tor Vergata. The discussions there with Professor L. ACCARDI, Professor P.A. MEYER and Professor K.R. PARTHASARATHY are gratefully acknowledged. For his interest in my project on indistinguishable particles and the kind hospitality in Rome I thank Professor L. ACCARDI.

2. Indistinguishable Quantum Particles

This chapter is a brief introduction into the fundamentals of the quantum theory of indistinguishable particles. This concept is defined here by means of the symmetry of the state (and not by the symmetry of observables). The results are confined to the essentials that are necessary or convenient for an understanding the classical part of the theory. The framework is elementary non relativistic quantum theory in an abstract Hilbert space setting. Therefore, we do not proceed to the Fock spaces nor do we consider scattering processes. This chapter is independent from the rest of this work.

In contrast to the familiar treatment of indistinguishable quantum particles, where besides BE statistics and FD statistics no quantum MB statistics is known, we introduce here a quantum generalization of classical MB statistics.

To maintain the text readable for physicists, we have tried to formulate the results as far as possible in terms of statistical operators and not in terms of states (in the sense of the theory of C*–algebras). Whenever, for notational convenience, properties are more easily expressed by states a notation such as $E_n(\cdot)$ should be read as $\mathrm{tr}(W_n \cdot)$.

We conclude these introductory remarks with some bibliographical information. As far as quantum theory is concerned we refer to the monographs by GALINDO and PASCUAL [75], JAUCH [95] and MESSIAH [116]. The necessary mathematical tools are explained, e.g., by REED and SIMON [134]. Quantum logics is developed, e.g., by BELTRAMETTI and CASSINELLI [28]. For quantum probability we refer to the lecture notes by MEYER [118] and the volume by PARTHASARATHY [125]. In particular, standard articles, concerned with the traditional viewpoint concerning indistinguishability are by MESSIAH and GREENBERG [117] and by HARTLE and TAYLOR [86].

2.1 Fundamentals

In this section we develop the fundamentals for the description of indistinguishable quantum particles. In particular, the definitions of the concept of identity and indistinguishability are presented. Moreover, pure symmetric states are analyzed. Finally, the three fundamental statistics are introduced.

2.1.1 Notation

Operators and states on $\mathcal{H}$. By a Hilbert space we understand a complex separable Hilbert space. Let be $\mathcal{H}$ a Hilbert space, then $\mathcal{B}(\mathcal{H})$ is the C*-algebra of bounded linear operators on $\mathcal{H}$, $\mathcal{B}_{sa}(\mathcal{H})$ is the subalgebra of self-adjoint elements, $\mathcal{P}(\mathcal{H})$ is the lattice of orthogonal projections, and $\mathcal{S}(\mathcal{H})$ is the convex set of statistical operators on $\mathcal{H}$. Moreover, $\mathcal{S}(\mathcal{B}(\mathcal{H}))$ is the set of states on $\mathcal{B}(\mathcal{H})$ and $\mathcal{S}(\mathcal{P}(\mathcal{H}))$ is the set of probability measures on $\mathcal{P}(\mathcal{H})$. In the context of the projection lattice $\mathcal{P}(\overset{n}{\otimes}\, \mathcal{H})$ we use the natural correspondence between a closed subspace of a Hilbert space and the orthogonal projection onto that subspace.

Assume that $W \in \mathcal{S}(\mathcal{H})$ and $P \in \mathcal{P}(\mathcal{H})$ are given such that $\mathrm{tr}(W\,P) \neq 0$, then the *conditional statistical operator* W_P of W, given P, is defined by (cf. e.g. [39])

$$W_P = \frac{P\,W\,P}{\mathrm{tr}(P\,W)}. \tag{2.1}$$

Obviously $W_P \in \mathcal{S}(\mathcal{H})$.

Tensor Products. Let $\mathcal{H}_1, \ldots, \mathcal{H}_n$ be Hilbert spaces and $\mathcal{H} = \mathcal{H}_1 \otimes \cdots \otimes \mathcal{H}_n$, then the *-algebra generated by $\{b_1 \otimes \cdots \otimes b_n; b_i \in \mathcal{B}(\mathcal{H}_i), 1 \leq i \leq n\}$ is strongly dense in $\mathcal{B}(\mathcal{H})$ such that $\overset{n}{\otimes}\, \mathcal{B}(\mathcal{H}) = \mathcal{B}(\overset{n}{\otimes}\, \mathcal{H})$ holds for any finite n. For any $W \in \mathcal{S}(\mathcal{H})$ and any $m, 1 \leq m \leq n$, there exists an uniquely determined statistical operator $W_{\hat{m}} \in \mathcal{S}(\mathcal{H}_1 \otimes \cdots \otimes \mathcal{H}_{\hat{m}} \otimes \cdots \mathcal{H}_n)$ – a caret means that a factor or component is omitted – such that

$$\mathrm{tr}(W_n\, b_1 \otimes \cdots \otimes b_{m-1} \otimes \mathbf{1} \otimes b_m \otimes \cdots \otimes b_{n-1}) = \mathrm{tr}(W_{\hat{m}}\, b_1 \otimes \cdots \otimes b_{n-1}) \tag{2.2}$$

holds for any $b_i \in \mathcal{B}(\mathcal{H}_i), 1 \leq i \leq n-1$. The map $R_{\hat{m}} : \mathcal{S}(\mathcal{H}_1 \otimes \cdots \otimes \mathcal{H}_m \otimes \cdots \otimes \mathcal{H}_n) \to \mathcal{S}(\mathcal{H}_1 \otimes \cdots \otimes \mathcal{H}_{\hat{m}} \otimes \cdots \otimes \mathcal{H}_n)$, defined by $R_{\hat{m}}(W) = W_{\hat{m}}$ is convex and called the *partial trace* with respect to the mth factor. $W_{\hat{m}}$ is called the *marginal statistical operator*.

A state defined in terms of a statistical operator $W \in \mathcal{S}(\mathcal{H})$ is called a *product state* if, $w_i \in \mathcal{S}(\mathcal{H}_i), 1 \leq i \leq n$,

$$W = w_1 \otimes \cdots \otimes w_n. \tag{2.3}$$

The particles described by a product state are called *statistically independent*. For $\mathcal{H}_i = \mathcal{H}_1, 1 \leq i \leq n$, a product state is called a *homogeneous product state* if $w_i = w_1, 1 \leq i \leq n$.

Permutation Operators. Let $\mathcal{H}$ be a Hilbert space and $\xi_i, i \in \mathbf{N}$, a complete orthonormal system of $\mathcal{H}$. Then the set of vectors

$$\xi_{i_1} \otimes \cdots \otimes \xi_{i_n}, \qquad (i_1, \ldots, i_n) \in \mathbf{N}^n$$

is a complete orthonormal system of the tensor product $\overset{n}{\otimes}\, \mathcal{H}$ of n identical copies of $\mathcal{H}$. The group of permutations of the integers $\{1, \ldots, n\}$ is denoted by S_n. For any $n \in \mathbf{N}$ and any $\pi \in S_n$ we denote by U_π the unitary *permutation operator* on $\overset{n}{\otimes} \mathcal{H}$ that is defined by its action on a complete orthonormal system

$$U_\pi \, \xi_{i_1} \otimes \cdots \otimes \xi_{i_n} = \xi_{i_{\pi(1)}} \otimes \cdots \otimes \xi_{i_{\pi(n)}} \tag{2.4}$$

and extended by linearity. The permutation operators define an unitary (in general reducible) representation of S_n since $U_{id} = \mathbf{1}, U_{\pi^{-1}} = U_\pi^+$ and $U_{\pi_1 \pi_2} = U_{\pi_1} U_{\pi_2}$ holds for any $\pi, \pi_1, \pi_2 \in S_n$.

For operators $b_1, \ldots, b_n$ that are defined on the one-particle Hilbert space $\mathcal{H}$ the operator $b_1 \otimes \cdots \otimes b_n$ which is defined on $\overset{n}{\otimes} \mathcal{H}$ transforms according to

$$U_\pi \, b_1 \otimes \cdots \otimes b_n \, U_\pi^+ = b_{\pi(1)} \otimes \cdots \otimes b_{\pi(n)}. \tag{2.5}$$

Symmetric Operators. An operator $B \in \mathcal{B}(\overset{n}{\otimes} \mathcal{H})$ is called *symmetric* if

$$U_\pi B U_\pi^+ = B \tag{2.6}$$

holds for all permutations $\pi \in S_n$. The subalgebra of all symmetric operators is denoted by $\mathcal{B}_{\text{sym}}(\overset{n}{\otimes} \mathcal{H})$ and the set of symmetric statistical operators on $\overset{n}{\otimes} \mathcal{H}$ is denoted by $\mathcal{S}_{\text{sym}}(\overset{n}{\otimes} \mathcal{H})$. By $\mathcal{P}_{\text{sym}}(\overset{n}{\otimes} \mathcal{H}) = \mathcal{P}(\overset{n}{\otimes} \mathcal{H}) \cap \mathcal{B}_{\text{sym}}(\overset{n}{\otimes} \mathcal{H})$ we denote the set of symmetric orthogonal projections.

Lemma 2.1.1. *Let be $P \in \mathcal{P}(\overset{n}{\otimes} \mathcal{H})$, then $P(\overset{n}{\otimes} \mathcal{H})$ is an invariant subspace under the action of S_n iff $P \in \mathcal{P}_{\text{sym}}(\overset{n}{\otimes} \mathcal{H})$.*

Proof. Recall that a subspace $P(\overset{n}{\otimes} \mathcal{H})$ of $\overset{n}{\otimes} \mathcal{H}$ is invariant if $\Phi = P\Phi$ implies $U_\pi \Phi = P U_\pi \Phi$ for any $\pi \in S_n$ and that the orthogonal complement $(\mathbf{1} - P)(\overset{n}{\otimes} \mathcal{H})$ of an invariant subspace $P(\overset{n}{\otimes} \mathcal{H})$ is invariant too.

Assume that $\Phi = P\Phi$ and that $[U_\pi, P] = 0$ for any $\pi \in S_n$. This yields $U_\pi \Phi = U_\pi P \Phi = P U_\pi \Phi$ for any $\pi \in S_n$ such that $P(\overset{n}{\otimes} \mathcal{H})$ is invariant. Conversely, assume that $P(\overset{n}{\otimes} \mathcal{H})$ is invariant. Then $(\mathbf{1} - P)(\overset{n}{\otimes} \mathcal{H})$ is invariant and it is sufficient to show that $[U_\pi, P] = 0$ holds on each of these subspaces. So suppose that $\Phi = P\Phi$. Invariance implies that $U_\pi \Phi = P U_\pi \Phi$. Replacing Φ by $P\Phi$ yields the assumption and the same strategy applies to $\Phi = (\mathbf{1} - P)\Phi$.
QED

Whenever $\Phi \in \overset{n}{\otimes} \mathcal{H}$ is an eigenvector of an operator $B \in \mathcal{B}_{\text{sym}}(\overset{n}{\otimes} \mathcal{H})$ with eigenvalue $\lambda \in \mathbf{C}$ the orthogonal projection onto the corresponding eigenspace is symmetric since $B\Phi = \lambda\Phi$ implies $B U_\pi \Phi = \lambda U_\pi \Phi$ for any $\pi \in S_n$ such that the eigenspace belonging to λ is the invariant subspace generated by Φ. Therefore, for any compact self-adjoint operator $B \in \mathcal{S}_{\text{sym}}(\overset{n}{\otimes} \mathcal{H})$ (and in particular for any symmetric statistical operator) the spectral representation

$$B = \sum_i b_i P_i \tag{2.7}$$

is defined in terms of symmetric orthogonal projections, i.e. $P_i \in \mathcal{P}_{\mathrm{sym}}(\overset{n}{\otimes} \mathcal{H})$.

The linear operator $E_{\mathrm{sym}} : \overset{n}{\otimes} \mathcal{B}(\mathcal{H}) \to \mathcal{B}_{\mathrm{sym}}(\overset{n}{\otimes} \mathcal{H})$ defined by

$$E_{\mathrm{sym}}(\,\cdot\,) = \frac{1}{n!} \sum_\pi U_\pi \cdot U_\pi^+, \tag{2.8}$$

is the quantum conditional expectation onto the subalgebra of symmetric operators, that is, E_{sym} is a norm one projection

$$E_{\mathrm{sym}}(E_{\mathrm{sym}}(\,\cdot\,)) = E_{\mathrm{sym}}(\,\cdot\,), \qquad E_{\mathrm{sym}}(\mathbf{1}) = \mathbf{1}. \tag{2.9}$$

Remark 2.1.1.

(a) The restriction of E_{sym} to the self–adjoint operators, positive operators respectively, is structure preserving. Moreover, the restriction of E_{sym} to the trace class operators is trace preserving. In particular, E_{sym} maps $\mathcal{S}(\overset{N}{\otimes} \mathcal{H})$ onto $\mathcal{S}_{\mathrm{sym}}(\overset{n}{\otimes} \mathcal{H})$

$$E_{\mathrm{sym}}(\mathcal{S}(\overset{n}{\otimes} \mathcal{H})) = \mathcal{S}_{\mathrm{sym}}(\overset{n}{\otimes} \mathcal{H}). \tag{2.10}$$

(b) On the other hand, for $P \in \mathcal{P}(\overset{n}{\otimes} \mathcal{H})$ in general $E_{\mathrm{sym}}(P) \notin \mathcal{P}_{\mathrm{sym}}(\overset{n}{\otimes} \mathcal{H})$, i.e.

$$E_{\mathrm{sym}}(\mathcal{P}(\overset{n}{\otimes} \mathcal{H})) \supset \mathcal{P}_{\mathrm{sym}}(\overset{n}{\otimes} \mathcal{H}). \tag{2.11}$$

As an example consider $Q_1 \otimes Q_2 \in \mathcal{P}(\overset{2}{\otimes} \mathcal{H})$ where $Q_i =< \xi_i, \cdot > \xi_i$ for two orthogonal unit vectors ξ_1, ξ_2. In this case $E_{\mathrm{sym}}(Q_1 \otimes Q_2) \notin \mathcal{P}_{\mathrm{sym}}(\overset{2}{\otimes} \mathcal{H})$.

Lemma 2.1.2. *Assume that $B \in \mathcal{B}(\overset{n}{\otimes} \mathcal{H})$ and that $W \in \mathcal{S}(\overset{n}{\otimes} \mathcal{H})$.*

1. *If B is symmetric*

$$\mathrm{tr}(W\,B) = \mathrm{tr}(E_{\mathrm{sym}}(W)\,B) \tag{2.12}$$

holds.

2. *If W is a symmetric statistical operator*

$$\mathrm{tr}(W\,B) = \mathrm{tr}(W\,E_{\mathrm{sym}}(B)) \tag{2.13}$$

holds.

Proof. (i) Since for any $\pi \in S_n$ the identity

$$\mathrm{tr}(W\,B) = \mathrm{tr}(W\,U_\pi\,B\,U_\pi^+) = \mathrm{tr}(U_\pi^+\,W\,U_\pi\,B) \tag{2.14}$$

holds we obtain

$$n!\,\mathrm{tr}(W\,B) = \mathrm{tr}(\sum_\pi U_\pi\,W\,U_\pi^+\,B). \tag{2.15}$$

For (ii) we use the same property. QED

2.1.2 Indistinguishability

Identity.

Definition 2.1.1. *(Cf. e.g. [95]) Particles are called* identical *if they agree in all their intrinsic (i.e. state independent) properties.*

Remark 2.1.2.
(a) The existence of identical particles is the fundamental assumption of all variants of atomism from antiquity until today.
(b) If the particles are described by a Hamiltonian H they are identical iff H is symmetric.

Indistinguishability.

Definition 2.1.2. *Identical particles are called* indistinguishable *if they are in a state that is symmetric.*

Remarks.

1. *Indistinguishability is a genuine probabilistic concept.* This is obvious in the quantum framework where the state is given by a statistical operator. Definition 2.1.2 applies to a classical statistical setting, where the state is determined by a probability measure such that indistinguishable classical particles are identical particles in a state that is determined by a symmetric probability measure. In a classical deterministic setting, however, no meaningful concept of indistinguishability exists since this implies that all particles are confined to the same point in configuration space or phase space which is in conflict with the impenetrability property of the (classical) particle concept (cf. Section 1.2).
2. *There are no indistinguishable particles.* Indistinguishability is a property of the state and not of the particles, i.e. indistinguishability – in the setting considered here (elementary non relativistic quantum theory) – is not an ontological property of certain objects but can be prepared or can be avoided by preparations (it may even be time–dependent). For a system of macroscopically many identical particles – e.g. for a gas –, however, it is empirically impossible to prepare a system in a state where the particles of one species are identifiable (e.g. by their positions).
The fact that indistinguishable particles are labeled (by a number) is a property of the description and not an empirical property. Permutation invariance means that the association: particle $\leftrightarrow$ number, is irrelevant.
If the kinematics is *a priori* restricted to a state space where only symmetric statistical operators or symmetric statistical operators of a certain type are admitted, it is obvious that in this restricted setting identity implies indistinguishability.
3. *Identity does not imply indistinguishability* (cf. e.g. [136]). If the statistical operator W is a function of a symmetric Hamiltonian H, then

identity implies indistinguishability. Therefore, identical particles are indistinguishable in the canonical (grand canonical) thermal equilibrium state of an ideal gas.
As an example for identical particles that are not indistinguishable we consider two identical particles which are described by the wave function (defined on the configuration space)

$$\Psi(\mathbf{x}_1, \mathbf{x}_2) = \Psi_1(\mathbf{x}_1)\Psi_2(\mathbf{x}_2), \tag{2.16}$$

where the supports of the functions Ψ_1 and Ψ_2 are disjoint. This implies that the supports of the associated marginal probability densities are disjoint such that these identical particles can be distinguished by their positions.
The fact that we do not need to anti/symmetrize the state of all identical particles of one species in the universe is usually considered in a spatial context where groups of identical particles are far apart from one another. This is called the *cluster law*. STEINMANN [137] and LÜDERS [112] have argued that the cluster law is compatible with bosons and fermions only. HARTLE and TAYLOR [86] have shown that the cluster law is compatible with parastatistics.
In our setting it is obvious that the cluster law is compatible with symmetry and is not restricted to certain types of symmetry. The factorization underlying the cluster law has nothing to do with a spatial separation but expresses statistical independence. Altogether we conclude that there exists no principle of indistinguishability of identical particles.

4. *Equivalence of the definition by means states and observables.* Usually (see e.g. [116, 95]) indistinguishability is defined by means of the symmetry of observables. It is obvious, and has been shown explicitly, e.g. by DRESDEN [54], that the assumption of symmetric observables does not imply that the observables are necessarily symmetric.
According to Lemma 2.1.2 it does not matter whether the theory of indistinguishable particles is founded on symmetric observables or symmetric states. In either case the dual quantity can be restricted to its symmetrized variant.
5. Founding the theory on symmetric states has the following advantages.
   - There is an obvious concept of classical indistinguishable particles.
   - The correlations for the particles of classical BE statistics and FD statistics are well defined. The particles of classical MB statistics are indistinguishable.
   - There exists a classification (extendibility see Section 2.2) that is more useful than the traditional one (representation theory of S_n).
   - One of the most powerful theoretical tools in this context, de Finetti's theorem (cf. Section 2.2), refers to the object that, by definition, carries the symmetry.
   - Sometimes, see e.g. Section 2.3, it is useful to consider explicitly observables that are not symmetric.

Marginals.

Lemma 2.1.3. *Assume that* $W_n \in S_{\text{sym}}(\overset{n}{\otimes}\mathcal{H})$, *then*

1. *for any* $m, 1 \leq m < n$, *the* $\binom{n}{m}$ *reduced statistical operators defined on* $\overset{m}{\otimes}\mathcal{H}$ *are identical and are, therefore, denoted by* W_m, *and*
2. *for any* $m, 2 \leq m < n$, *the statistical operators* W_m *are symmetric.*

Proof. It is sufficient to show, that for any $m, 1 \leq m \leq n$, the reduced statistical operator $W_{\hat{m}}$ (partial trace over the mth factor of the tensor product) defined on $\overset{n-1}{\otimes}\mathcal{H}$ is symmetric and that for $m_1 \neq m_2$ we have $W_{\hat{m}_1} = W_{\hat{m}_2}$. Then we can proceed by induction.

(i) Since W_n is symmetric under S_n, for the permutation $\pi \in S_n$ that exchanges particle at site m_1 with particle at site m_2 we obtain

$$\begin{aligned}
&\operatorname{tr}(W_{\hat{m}_1}\, b_1 \otimes \ldots \otimes b_{n-1}) \\
&= \operatorname{tr}(W_n\, b_1 \otimes \cdots \otimes b_{m_1-1} \otimes \mathbf{1} \otimes b_{m_1} \otimes \cdots \otimes b_{n-1}) \\
&= \operatorname{tr}(U_\pi\, W_n\, U_\pi^+\, b_1 \otimes \cdots \otimes b_{m_1-1} \otimes \mathbf{1} \otimes b_{m_1} \otimes \cdots \otimes b_{n-1}) \\
&= \operatorname{tr}(W_n\, U_\pi^+\, b_1 \otimes \cdots \otimes b_{m_1-1} \otimes \mathbf{1} \otimes b_{m_1} \otimes \cdots \otimes b_{n-1}\, U_\pi) \\
&= \operatorname{tr}(W_n\, b_1 \otimes \cdots \otimes b_{m_2-1} \otimes \mathbf{1} \otimes b_{m_2} \otimes \cdots \otimes b_{n-1}) \\
&= \operatorname{tr}(W_{\hat{m}_2}\, b_1 \otimes \cdots \otimes b_{n-1}).
\end{aligned} \tag{2.17}$$

This implies $W_{\hat{m}_1} = W_{\hat{m}_2}$ since the observables separate the states. (ii) Assume that $W_{\hat{m}}$ is not symmetric. Then W_n is not symmetric since it is not symmetric under the subgroup $S_n^{(m)}$ of S_n where the mth component is fixed.

QED

2.1.3 Pure Symmetric States

Symmetric and Antisymmetric Vectors. The operators $\Pi_+, \Pi_- \in \overset{n}{\otimes} \mathcal{B}(\mathcal{H})$ defined by

$$\Pi_+ = \frac{1}{n!}\sum_\pi U_\pi, \quad \Pi_- = \frac{1}{n!}\sum_\pi \operatorname{sgn}(\pi) U_\pi \tag{2.18}$$

where $\operatorname{sgn}(\pi) \in \{-1, 1\}$ denotes the sign of the permutation π are orthogonal projections onto mutually orthogonal subspaces of $\overset{n}{\otimes}\mathcal{H}$ that satisfy for any $\pi \in S_n$

$$U_\pi\, \Pi_+ = \Pi_+, \qquad U_\pi\, \Pi_- = \operatorname{sgn}(\pi)\, \Pi_-. \tag{2.19}$$

Therefore $\Pi_\pm$ are invariant under the action of the symmetric group, i.e. $\Pi_+, \Pi_- \in \mathcal{P}_{\text{sym}}(\overset{n}{\otimes}\mathcal{H})$. The subspace $(\overset{n}{\otimes}\mathcal{H})_+ = \Pi_+(\overset{n}{\otimes}\mathcal{H})$ is called the BE symmetric subspace and its elements are called *symmetric vectors*. The subspace $(\overset{n}{\otimes}\mathcal{H})_- = \Pi_-(\overset{n}{\otimes}\mathcal{H})$ is called the FD symmetric subspace and its elements are called *antisymmetric vectors*. Notice that for two particles $n=2$ the identity

$$\text{for } n=2: \qquad \Pi_+ + \Pi_- = \mathbf{1} \tag{2.20}$$

holds such that in this particular case the total Hilbert space is the direct sum of the BE symmetric subspace and the FD symmetric subspace.

The next lemma shows that the set of homogeneous product vectors is a total set in $\Pi_+(\overset{n}{\otimes}\mathcal{H})$. (A set $\mathcal{K}$ in a Hilbert space $\mathcal{H}$ is total if its closed linear span is $\mathcal{H}$. A total set that is strictly larger than a c.o.n.s. is called overcomplete.)

Lemma 2.1.4. *(Cf. [5]) Assume that $\xi_1, \ldots \xi_n \in \mathcal{H}$ then*

$$\Pi_+ \xi_1 \otimes \cdots \otimes \xi_n = \frac{1}{2^n\, n!} \sum_{\underline{\epsilon}} \Big(\prod_{i=1}^{n} \epsilon_i\Big)\Big\{\overset{n}{\otimes}\Big(\sum_{i=1}^{n} \epsilon_i \xi_i\Big)\Big\}, \tag{2.21}$$

where the sum over $\underline{\epsilon}$ extends over all $\underline{\epsilon} \in \{f : \{1, \ldots, n\} \to \{-1, 1\}\}$.

Theorem.

Theorem 2.1.1. *Assume that $W \in S_{\mathrm{sym}}(\overset{n}{\otimes}\mathcal{H})$ is pure such that $W = < \Phi, \cdot > \Phi$ for some $\Phi \in \overset{n}{\otimes}\mathcal{H}, |\Phi| = 1$, then Φ is either symmetric or antisymmetric.*

Proof. (Cf. e.g. [78].) Assume that $W =< \Phi, \cdot > \Phi$ holds. Symmetry entails that

$$< \Phi, \cdot > \Phi =< \Phi, U_\pi^+ \cdot > U_\pi \Phi \tag{2.22}$$

holds for all $\pi \in S_n$. Accordingly,

$$U_\pi \Phi =< \Phi, U_\pi \Phi > \Phi. \tag{2.23}$$

Setting

$$c(\pi, \Phi) =< \Phi, U_\pi \Phi >, \quad c(id, \Phi) = 1, \tag{2.24}$$

the composition law of permutation operators, $U_{\pi_2 \pi_1} = U_{\pi_2} U_{\pi_1}$, entails

$$c(\pi_2 \pi_1, \Phi) = c(\pi_2, \Phi) c(\pi_1, \Phi), \tag{2.25}$$

such that $c(\pi, \Phi)$ is a one–dimensional representation of the symmetric group. For a transposition τ we have $U_\tau = U_\tau^+$ such that

$$c^2(\tau, \Phi) = 1. \tag{2.26}$$

Since any permutation is a product of an uniquely determined minimal number of mutually different transpositions, eventually this yields either

$$U_\pi \Phi = \Phi \tag{2.27}$$

for all $\pi \in S_n$ or

$$U_\pi \Phi = \text{sgn}(\pi)\Phi \tag{2.28}$$

for all $\pi \in S_n$. Since $\Psi \in \overset{n}{\otimes} \mathcal{H}$ is symmetric (antisymmetric) iff

$$U_\pi \Psi = \Psi, \quad (U_\pi \Psi = \text{sgn}(\pi)\Psi) \tag{2.29}$$

holds for all $\pi \in S_n$, this ends the proof. QED

Definition 2.1.3. *A pure symmetric statistical operator* $W = <\Phi, \cdot> \Phi$ *is called (i) a* pure BE symmetric statistical operator *if* Φ *is symmetric and (ii) a* pure FD symmetric statistical operator *if* Φ *is antisymmetric.*

Remark: Fractal Statistics. If we replace the abstract Hilbert space setting – $\overset{n}{\otimes} \mathcal{H}$ – by a setting involving more structure, e.g. a configuration space $\overset{n}{\times} X$, $\mathcal{H} = L^2(X, \mu), \overset{n}{\otimes} \mathcal{H} = L^2(\overset{n}{\times} X, \overset{n}{\times} \mu)$, because of the richer structure new phenomena appear (cf. e.g. [78]).

First, to account for the impenetrability of the particles in $X = \mathbf{R}^d, 1 \leq d \leq 3$, it is necessary the remove the 'diagonal' $D_n = \{(\mathbf{x}_1, \ldots, \mathbf{x}_n) \in \mathbf{R}^{dn}$; there are $i, j, 1 \leq i, j \leq n, i \neq j$, such that $\mathbf{x}_i = \mathbf{x}_j\}$ which yields the phase space $\mathbf{R}^{dn} \setminus D_n$. Second, to account for indistinguishability on can – in advance – restrict the kinematics, i.e. restrict this phase space to its symmetric quotient set such that one obtains the phase space [108]

$$\mathcal{M}_{n,d} = (\mathbf{R}^{dn} \setminus D_n)/S_n. \tag{2.30}$$

For $d = 3$ this phase space is simply connected, whereas for $d = 1$ and $d = 2$ it is multiply connected. In the latter case (cf. e.g. [68]) the associated first homotopy group (fundamental group) is the *braid group* of order n and the states induced by state vectors of this type are called *fractal statistics.*

2.1.4 Statistics

Definitions.

Definition 2.1.4. *Indistinguishable particles are called*

1. particles of quantum Maxwell-Boltzmann statistics *if the statistical operator* W *is* Maxwell-Boltzmann symmetric, *that is,* W *is a homogeneous product state*
$$W = \overset{n}{\otimes} w, \tag{2.31}$$
where w *is a statistical operator on* $\mathcal{H}$,
2. bosons *if the statistical operator* W *is* Bose-Einstein symmetric, *that is,*
$$U_\pi W = W \tag{2.32}$$
holds for all $\pi \in S_n$,
3. fermions *if the statistical operator* W *is* Fermi-Dirac symmetric, *that is,*
$$U_\pi W = \text{sgn}(\pi)\, W \tag{2.33}$$
holds for all $\pi \in S_n$.

Remarks.

Remark 2.1.3.

(a) Obviously any MB symmetric, any BE symmetric and any FD symmetric statistical operator is symmetric.

(b) The Definition 2.1.4 of BE/FD symmetric states is consistent with Definition 2.1.3 of pure BE/FD symmetric states. (i) Pure BE/FD symmetric states (in the sense of Definition 2.1.3) are BE/FD symmetric. (ii) Any mixture of pure BE/FD symmetric states is BE/FD symmetric.

(c) In the spectral representation of BE/FD symmetric statistical operators $W = \sum_i p_i P_i, P_i \in \mathcal{P}(\overset{n}{\otimes} \mathcal{H})$ the elements of the spectral measure are BE/FD symmetric, i.e. $\Pi_\pm P_i = P_i$ (since the spectral representation is an orthogonal decomposition).

Lemma 2.1.5. *For the four pairs of assertions*

$$1. \quad U_\pi W = W, \quad \text{or} \quad U_\pi W = \text{sgn}(\pi) W \quad \text{for all} \quad \pi \in S_n. \tag{2.34}$$

$$2. \quad \Pi_\pm W = W. \tag{2.35}$$

$$3. \quad \Pi_\pm W \Pi_\pm = W. \tag{2.36}$$

$$4. \quad [\Pi_\pm , W] = 0. \tag{2.37}$$

the following implications separately for BE/FD statistics hold: $1 \Longleftrightarrow 2, 2 \Longleftrightarrow 3, 3 \not\Longleftrightarrow 4$.

Proof. 1. ($1 \Rightarrow 2$): This follows from the definition of $\Pi_\pm$.

2. ($2 \Rightarrow 1$): This follows from eq.(2.19).
3. ($2 \Rightarrow 3$): This is obvious.
4. ($3 \Rightarrow 2$): Eq.(2.36) implies

$$\text{tr}(\Pi_\pm W) = 1, \tag{2.38}$$

such that (cf. [93]) $\Pi_\pm W = W$ follows.
5. ($2 \Rightarrow 4$): The self–adjointness of W implies $\Pi_\pm W = W = W^+ = W \Pi_\pm$.
6. ($4 \not\Leftarrow 2$): Use, as a counterexample $W = \frac{1}{2} W_{BE} + \frac{1}{2} W_{FD}$, where W_{BE} is BE symmetric and W_{FD} is FD symmetric. We have $[\Pi_\pm, W] = 0$ but $\Pi_\pm W \neq W$.

QED

Marginals.

Lemma 2.1.6. *Assume that W_n is MB symmetric, BE symmetric, FD symmetric, then for any $m, 2 \leq m < n$, the reduced statistical operator W_m defined on $\overset{m}{\otimes} \mathcal{H}$ is MB symmetric, BE symmetric, FD symmetric.*

Proof. For MB statistics the assertion is obvious. For BE and FD statistics it is sufficient to show that W_{n-1} has the same symmetry type as W_n and to proceed by induction.

For FD statistics we have to show that for any $b_1, \ldots b_{n-1} \in \mathcal{B}(\mathcal{H})$ and any $\pi \in S_{n-1}$ the identity

$$\mathrm{tr}(U_\pi W_{n-1} b_1 \otimes \cdots \otimes b_{n-1}) = \mathrm{tr}(\mathrm{sgn}(\pi) W_{n-1} b_1 \otimes \cdots \otimes b_{n-1}) \tag{2.39}$$

holds. By $\pi_n \in S_n$ we denote the embedding of π in S_n, i.e. $\pi_n(k) = \pi(k)$ for $k < n$ and $\pi_n(n) = n$. Notice that $\mathrm{sgn}(\pi) = \mathrm{sgn}(\pi_n)$ since the number of minimal transpositions is equal.

Using the identity

$$\{b_1 \otimes \cdots \otimes b_{n-1} U_\pi\} \otimes \mathbf{1} = \{b_1 \otimes \cdots \otimes b_{n-1} \otimes \mathbf{1}\} U_{\pi_n} \tag{2.40}$$

we obtain

$$\begin{aligned}
&\mathrm{tr}(U_\pi W_{n-1} b_1 \otimes \cdots \otimes b_{n-1}) \\
&= \mathrm{tr}(W_{n-1} b_1 \otimes \cdots \otimes b_{n-1} U_\pi) = \mathrm{tr}(W_n \{b_1 \otimes \cdots \otimes b_{n-1} U_\pi\} \otimes \mathbf{1}) \\
&= \mathrm{tr}(W_n \{b_1 \otimes \cdots \otimes b_{n-1} \otimes \mathbf{1}\} U_{\pi_n}) = \mathrm{tr}(U_{\pi_n} W_n b_1 \otimes \cdots \otimes b_{n-1} \otimes \mathbf{1}) \\
&= \mathrm{tr}(\mathrm{sgn}(\pi_n) W_n b_1 \otimes \cdots \otimes b_{n-1} \otimes \mathbf{1}) \\
&= \mathrm{tr}(\mathrm{sgn}(\pi) W_{n-1} b_1 \otimes \cdots \otimes b_{n-1}).
\end{aligned} \tag{2.41}$$

This implies $U_\pi W_{n-1} = \mathrm{sgn}(\pi) W_{n-1}$ for any $\pi \in S_{n-1}$. For BE statistics we apply the same strategy. QED

Remark 2.1.4. Since there exists at least one more symmetry type (mixtures of mixed MB symmetric states) that is invariant under marginalization the lemma cannot be considered as a characterization of the three statistics.

Problem 2.1.1. Do there exist more symmetry types than BE/FD symmetric states and mixtures of MB symmetric states that are invariant under marginalization?

Intersections.

Lemma 2.1.7. *[18] Denote by MB, BE, FD the sets of MB symmetric, BE symmetric and FD symmetric statistical operators on $\overset{n}{\otimes} \mathcal{H}$, then*

1. $BE \cap FD = \emptyset$,
2. $MB \cap FD = \emptyset$,
3. *$W \in MB \cap BE$ iff W is MB symmetric and pure and this holds iff $W = \overset{n}{\otimes} w$ and w is pure.*

Proof. (i) This is obvious since $\Pi_+ \Pi_- = \Pi_- \Pi_+ = 0$. For (ii) and (iii) observe that if $W = \overset{n}{\otimes} w$ is MB symmetric then the 2-dimensional reduced statistical operator $W_2 = w \otimes w$ is MB symmetric. Denoting by $\tau \in S_2$ the transposition that permutes particle 1 and 2, we evaluate the trace of $U_\tau w \otimes w$

$$\mathrm{tr}(U_\tau w \otimes w) = \sum_{i,j} < \xi_i \otimes \xi_j, w \otimes w \xi_j \otimes \xi_i > = \sum_{i,j} < \xi_i, w \xi_j > < \xi_j, w \xi_i >, \tag{2.42}$$

such that

$$\mathrm{tr}(U_\tau\, w \otimes w) \;=\; \mathrm{tr}(w^2). \tag{2.43}$$

(ii) Assume that W is a homogeneous product state on $\overset{n}{\otimes}\,\mathcal{H}$ and that W is FD-symmetric, then the 2-dimensional reduced statistical operator is FD-symmetric so that in particular for the transposition τ

$$U_\tau\, w \otimes w \;=\; -\, w \otimes w \tag{2.44}$$

holds. Taking the trace on both sides yields the contradiction

$$-\,1 \;=\; \mathrm{tr}(-w \otimes w) \;=\; \mathrm{tr}(U_\tau\, w \otimes w) \;=\; \mathrm{tr}(w^2) \;>\; 0. \tag{2.45}$$

(iii) First observe for a MB symmetric W that $W = \overset{n}{\otimes}\, w$ is pure if and only if w is pure. Therefore, if W is MB symmetric and pure, $W =<\Psi, \cdot > \Psi$, the product structure and homogeneity entail $\Psi = \overset{n}{\otimes}\, \psi$. This vector is BE symmetric and therefore the associated state is BE symmetric. Conversely, assume that the homogeneous product state W is BE symmetric, then the 2–dimensional reduced statistical operator is BE symmetric so that in particular for the transposition τ

$$U_\tau\, w \otimes w \;=\; w \otimes w. \tag{2.46}$$

holds. Taking the trace on both sides gives

$$(\mathrm{tr}(w))^2 \;=\; \mathrm{tr}(w \otimes w) \;=\; \mathrm{tr}(U_\tau\, w \otimes w) \;=\; \mathrm{tr}(w^2). \tag{2.47}$$

This implies

$$1 \;=\; (\mathrm{tr}(w))^2 \;=\; \mathrm{tr}(w^2), \tag{2.48}$$

such that w is pure. QED

Remark 2.1.5. The states that are both MB symmetric and BE symmetric are the most elementary states of indistinguishable particles and constitute the prototype of coherent states. Examples of these distinguished states (independent bosons) are discussed in Section 2.3.

The Exclusion Principle.

Definition 2.1.5. *Assume that $W \in \mathcal{S}_{\mathrm{sym}}(\overset{n}{\otimes}\,\mathcal{H})$ and suppose that for any c.o.n.s. $\xi_i, i \le \dim(\mathcal{H})$ of $\mathcal{H}$ the relation*

$$< \xi_{i_1} \otimes \cdots \otimes \xi_{i_n}, W \xi_{i_1} \otimes \cdots \otimes \xi_{i_n} >= 0 \tag{2.49}$$

holds whenever there exist indices $k, j, 1 \le k, j \le n, j \neq k$, such that $i_j = i_k$, then W is said to satisfy the exclusion principle.

Lemma 2.1.8. *The exclusion property is invariant under marginalization.*

Proof. It is sufficient to show that for $W_{n-1} \in \mathcal{S}_{\text{sym}}(\overset{n-1}{\otimes} \mathcal{H})$ the exclusion principle holds if it holds for $W_n \in \mathcal{S}_{\text{sym}}(\overset{n}{\otimes} \mathcal{H})$. Assume that $\xi_i, i \leq \dim(\mathcal{H})$, is any c.o.n.s. of $\mathcal{H}$ and suppose that $1 \leq j, k \leq n-1$ and $j \neq k, i_j = i_k$. We have

$$
\begin{aligned}
&< \xi_{i_1} \otimes \cdots \otimes \xi_{i_{n-1}}, W_{n-1} \xi_{i_1} \otimes \cdots \otimes \xi_{i_{n-1}} > \\
&= \sum_{\ell} < \xi_{i_1} \otimes \cdots \otimes \xi_{i_{n-1}} \otimes \xi_\ell, W \xi_{i_1} \otimes \cdots \otimes \xi_{i_{n-1}} \otimes \xi_\ell > = 0.
\end{aligned} \tag{2.50}
$$

QED

Lemma 2.1.9.

1. *Any MB symmetric statistical operator violates the exclusion principle.*
2. *Any BE symmetric statistical operator violates the exclusion principle.*
3. *Any FD symmetric statistical operator satisfies the exclusion principle.*

Proof. 1. Assume that $W = \overset{n}{\otimes} w$ is MB symmetric and satisfies the exclusion principle, then

$$(< \xi, w\xi >)^n = 0 \quad \Rightarrow \quad < \xi, w\xi > = 0 \tag{2.51}$$

for any eigenvector ξ of w. Therefore the eigenvalues of w are all equal to zero which is impossible since w is a statistical operator.

2. Assume that W is BE symmetric and satisfies the exclusion principle, then in particular

$$< \xi_i \otimes \cdots \otimes \xi_i, W \xi_i \otimes \cdots \otimes \xi_i > = 0 \tag{2.52}$$

for any vector ξ_i that is contained in any c.o.n.s. Therefore

$$W \phi \otimes \cdots \otimes \phi = 0 \tag{2.53}$$

for any $\phi \in \mathcal{H}$. According to eq.(2.21) the homogeneous product vectors are a total set in $\Pi_+(\overset{n}{\otimes} \mathcal{H})$ such that, by BE symmetry, $W = 0$ follows in contradiction to the normalization of W.

3. Consider any c.o.n..s. $\xi_i, i \leq \dim(\mathcal{H})$, of $\mathcal{H}$ and a state vector $\xi_{i_1} \otimes \cdots \otimes \xi_{i_n}$ where $i_j = i_k, i \neq k$. This vector is invariant under the transposition U_τ that exchanges particle j and particle k. Assume that W is FD symmetric, then $W U_\tau = -W$ and we have

$$
\begin{aligned}
&< \xi_{i_1} \otimes \cdots \otimes \xi_{i_n}, W \xi_{i_1} \otimes \cdots \otimes \xi_{i_n} > \\
&= - < \xi_{i_1} \otimes \cdots \otimes \xi_{i_n}, W \xi_{i_1} \otimes \cdots \otimes \xi_{i_n} > .
\end{aligned} \tag{2.54}
$$

QED

Remark 2.1.6. There do exist BE symmetric states for which any multiple occupations in a *fixed* one–particle basis $\xi_i, 1 \leq i \leq \dim(\mathcal{H})$, have probability zero. Cf. e.g. the statistical operator defined by the state vector

$$\Phi_{BE} = \sqrt{n!}\, \Pi_+ \, \xi_{i_1} \otimes \cdots \otimes \xi_{i_n}, \tag{2.55}$$

where all indices are different.

2.2 Classification: Extremality and Extendibility

Traditionally, based upon the pioneering work by WEYL [153], indistinguishable particles (defined by the symmetry of their observables) are classified according to the representation theory of the symmetric group by invariant irreducible subspaces. In contrast to this algebraic procedure we use here probabilistic criteria for a classification. It turns out, however, that the first probabilistic classification according to extreme points is equivalent to the traditional algebraic classification. The second probabilistic classification is based upon extendibility properties and allows for an inner classification of the familiar two statistics.

2.2.1 Extreme Symmetric States

Theorem. For a fixed Hilbert space $\mathcal{H}$ and a fixed number n of particles the set of symmetric statistical operators $\mathcal{S}_{\text{sym}}(\overset{n}{\otimes}\mathcal{H})$ is a convex set such that the extreme points are of particular importance. In what follows by 'invariant' we shall always understand 'invariant under the action of the symmetric group'.

Theorem 2.2.1. *[19] $W \in \text{ex}(\mathcal{S}_{\text{sym}}(\overset{n}{\otimes}\mathcal{H}))$ iff $W = P/\text{tr}(P)$ for some orthogonal projection P onto a subspace of $\overset{n}{\otimes}\mathcal{H}$ that is invariant and irreducible.*

To motivate our proof, recall (cf. e.g. [28]) that there exists a one–to–one correspondence between the atoms of lattice of orthogonal projections $\mathcal{P}(\overset{n}{\otimes}\mathcal{H})$ and the extreme points of set of probability measures on this lattice which can be identified with the set $\mathcal{S}(\overset{n}{\otimes}\mathcal{H})$ (the extreme points with the set $\text{ex}(\mathcal{S}(\overset{n}{\otimes}\mathcal{H}))$), the correspondence being given by

$$W(P) = P, \qquad P(W) = W, \tag{2.56}$$

where $W \in \text{ex}(\mathcal{S}(\overset{n}{\otimes}\mathcal{H}))$ (W is a pure state) and $P \in \mathcal{P}(\overset{n}{\otimes}\mathcal{H})$ is an atom (a one–dimensional orthogonal projection). In the following we show an analogous correspondence for the sublattice $\mathcal{P}_{\text{sym}}(\overset{n}{\otimes}\mathcal{H})$ of orthogonal projections onto invariant subspaces. Let us start the proof with two statements that allow for a slight reformulation of the theorem.

Lemma 2.2.1. *$\mathcal{P}_{\text{sym}}(\overset{n}{\otimes}\mathcal{H})$ is an atomistic sublattice of $\mathcal{P}(\overset{n}{\otimes}\mathcal{H})$ and its atoms are the orthogonal projections onto irreducible invariant subspaces.*

Proof. By definition orthogonal projections onto irreducible invariant subspaces are the atoms. Recall that a lattice is atomic if it contains atoms, and that an atomic lattice is atomistic if any element is equal to the joint of (all) its atoms. Since $\mathcal{S}_{\text{sym}}(\overset{n}{\otimes}\mathcal{H})$ is a sublattice of the atomistic lattice $\mathcal{S}(\overset{n}{\otimes}\mathcal{H})$ it is atomistic. QED

By $\mathcal{S}(\mathcal{P}_{\mathrm{sym}}(\overset{n}{\otimes}\mathcal{H}))$ we denote the convex set of probability measures on $\mathcal{P}_{\mathrm{sym}}(\overset{n}{\otimes}\mathcal{H})$.

Lemma 2.2.2. *$\mathcal{S}_{\mathrm{sym}}(\overset{n}{\otimes}\mathcal{H})$ can be identified with a subset of $\mathcal{S}(\mathcal{P}_{\mathrm{sym}}(\overset{n}{\otimes}\mathcal{H}))$.*

Proof. Assume that $W \in \mathcal{S}_{\mathrm{sym}}(\overset{n}{\otimes}\mathcal{H})$. Obviously the prescription $\Pi : \mathcal{P}_{\mathrm{sym}}(\overset{n}{\otimes}\mathcal{H}) \to [0,1]$ defined by $\Pi(P) = \mathrm{tr}(W\,P)$ determines a probability measure on $\mathcal{P}_{\mathrm{sym}}(\overset{n}{\otimes}\mathcal{H})$. QED

The results derived so far admit the following reformulation of Theorem (2.2.1).

Theorem 2.2.2. *W is an extreme point of the convex set $\mathcal{S}_{\mathrm{sym}}(\overset{n}{\otimes}\mathcal{H})$ iff $W = P/\mathrm{tr}(P)$ for some atom P of $\mathcal{P}_{\mathrm{sym}}(\overset{n}{\otimes}\mathcal{H})$.*

For a proof of this theorem we need more information concerning the structure of the lattice $\mathcal{P}_{\mathrm{sym}}(\overset{n}{\otimes}\mathcal{H})$ and the subset $\mathcal{S}_{\mathrm{sym}}(\overset{n}{\otimes}\mathcal{H})$ of probability measures on $\mathcal{P}_{\mathrm{sym}}(\overset{n}{\otimes}\mathcal{H})$.

Lemma 2.2.3. *The elements of the lattice $\mathcal{P}_{\mathrm{sym}}(\overset{n}{\otimes}\mathcal{H})$ separate the probability measures $\mathcal{S}_{\mathrm{sym}}(\overset{n}{\otimes}\mathcal{H})$, i.e. $\mathrm{tr}(W_1P) = \mathrm{tr}(W_2P)$ for all $P \in \mathcal{P}_{\mathrm{sym}}(\overset{n}{\otimes}\mathcal{H})$ implies $W_1 = W_2$.*

Proof. Assume that $W_1, W_2 \in \mathcal{S}_{\mathrm{sym}}(\overset{n}{\otimes}\mathcal{H})$ and that $\mathrm{tr}(W_1P) = \mathrm{tr}(W_2P)$ for all $P \in \mathcal{P}_{\mathrm{sym}}(\overset{n}{\otimes}\mathcal{H})$. For $W \in \mathcal{S}_{\mathrm{sym}}(\overset{n}{\otimes}\mathcal{H})$ and $P \in \mathcal{P}(\overset{n}{\otimes}\mathcal{H})$ we obtain

$$\mathrm{tr}(WP) = \mathrm{tr}(WE_{\mathrm{sym}}(P)) = \sum_i q_i\,\mathrm{tr}(WQ_i), \tag{2.57}$$

where $E_{\mathrm{sym}}(P) = \sum_i q_iQ_i, q_i \in \mathbf{R}, Q_i \in \mathcal{P}_{\mathrm{sym}}(\overset{n}{\otimes}\mathcal{H})$ is the spectral decomposition ($E_{\mathrm{sym}}(P)$ is compact since P is so). Therefore

$$\begin{aligned}\mathrm{tr}(W_1P) = \mathrm{tr}(W_1E_{\mathrm{sym}}(P)) &= \sum_i q_i\,\mathrm{tr}(W_1Q_i) = \sum_i q_i\,\mathrm{tr}(W_2Q_i)\\ = \mathrm{tr}(W_2E_{\mathrm{sym}}(P)) &= \mathrm{tr}(W_2P)\end{aligned} \tag{2.58}$$

for any $P \in \mathcal{P}(\overset{n}{\otimes}\mathcal{H})$. Since the elements of the lattice $\mathcal{P}(\overset{n}{\otimes}\mathcal{H})$ separate the probability measures $\mathcal{S}(\overset{n}{\otimes}\mathcal{H})$ the assertion follows. QED

Lemma 2.2.4. *The set of probability measures $\mathcal{S}_{\mathrm{sym}}(\overset{n}{\otimes}\mathcal{H})$ is sufficient, i.e. for any $P \in \mathcal{S}_{\mathrm{sym}}(\overset{n}{\otimes}\mathcal{H}), P \neq 0$, there is a $W \in \mathcal{S}_{\mathrm{sym}}(\overset{n}{\otimes}\mathcal{H})$ such that $W(P) = 1$.*

Proof. Assume that $P \in \mathcal{P}_{\text{sym}}(\overset{n}{\otimes}\mathcal{H})$. Since $P \in \mathcal{P}(\overset{n}{\otimes}\mathcal{H})$ and the set of probability measures $\mathcal{S}(\overset{n}{\otimes}\mathcal{H})$ is sufficient, there exists a $W \in \mathcal{S}(\overset{n}{\otimes}\mathcal{H})$ such that $\text{tr}(WP) = 1$. From this $\text{tr}(E_{\text{sym}}(W)P) = 1$ follows and $E_{\text{sym}}(W)$ satisfies the condition. QED

Lemma 2.2.5. *For all* $P \in \mathcal{P}_{\text{sym}}(\overset{n}{\otimes}\mathcal{H}), P \neq 0$, *there exists an uniquely determined* $W_P \in \mathcal{S}_{\text{sym}}(\overset{n}{\otimes}\mathcal{H})$ *such that* $\text{tr}(W_P P) = 1$ *holds, namely* $W_P = P/\text{tr}(P)$.

Proof. Existence follows from Lemma 2.2.4, uniqueness from Lemma 2.2.3. By inspection it is clear that $\text{tr}(W_P P) = 1$ holds. QED

We introduce the following notation: ran(·) denotes the range of a mapping, lin(·) denotes the linear span of a set of vectors, cl(·) denotes the closure of a set in a given topology (here the strong topology of Hilbert spaces).

Lemma 2.2.6. *Assume that* T_1, T_2 *are positive semidefinite trace class operators on a Hilbert space* $\mathcal{H}$, *then* $\text{ran}(T_1 + T_2) = \text{cl}(\text{lin}(\text{ran}(T_1), \text{ran}(T_2)))$.

Proof. Since T_1, T_2 are positive semidefinite trace class operators, T_i admits the spectral representation, $i = 1, 2$,

$$T_i = \sum_j t_j^{(i)} < e_j^{(i)}, \cdot > e_j^{(i)}, \tag{2.59}$$

where $t_j^{(i)} > 0$ and $e_j^{(i)}$ are two orthonormal systems which are either complete or can be enlarged to a complete orthonormal system of $\mathcal{H}$. From this representation the assertion follows immediately. QED

We are prepared now to perform the proof of Theorem 2.2.1.

Proof. (i) Assume that $W = P/\text{tr}(P)$ and that $W = \lambda W_1 + (1-\lambda)W_2$ holds, where $W_1, W_2 \in \mathcal{S}_{\text{sym}}(\overset{n}{\otimes}\mathcal{H})$ and $\lambda \in (0,1)$. According to Lemma 2.2.6 $\text{ran}(W) = \text{cl}(\text{lin}(\text{ran}(W_1), \text{ran}(W_2)))$ is irreducible. This is possible only if $\text{ran}(W_1) = \text{ran}(W_2)$ and if $\text{ran}(W_1)$ is irreducible. We conclude $W_1 = W_2 = W$, such that W is extreme.

(ii) Assume that $W \in \mathcal{S}_{\text{sym}}(\overset{n}{\otimes}\mathcal{H})$ and that W is not extreme such that $W = \lambda W_1 + (1-\lambda)W_2$ holds where $W_1, W_2 \in \mathcal{S}_{\text{sym}}(\overset{n}{\otimes}\mathcal{H})$ and $\lambda \in (0,1)$ and $W_1 \neq W_2$. According to Lemma 2.2.6 $\text{ran}(W) = \text{cl}(\text{lin}(\text{ran}(W_1), \text{ran}(W_2)))$. If $\text{ran}(W_1)$ or $\text{ran}(W_2)$ are reducible, then $\text{ran}(W)$ is reducible. If both $\text{ran}(W_1)$ and $\text{ran}(W_2)$ are irreducible then $W_i = P_i/\text{tr}(P_i)$, $i = 1, 2$ and $P_1 \neq P_2$ so that $\text{ran}(P_1) \neq \text{ran}(P_2)$ which implies that $\text{ran}(W) = \text{cl}(\text{lin}(\text{ran}(P_1), \text{ran}(P_2)))$ is reducible. QED

Remarks.

Remark 2.2.1.

(i) The traditional algebraic classification and the probabilistic classification by means of extreme elements are equivalent.

(ii) Recall that the support of a probability measure on $\mathcal{P}(\overset{n}{\otimes}\mathcal{H})$ is given by the range of the associated statistical operator. The fundamental result of this section can be expressed in the following form. There exists a one-to-one correspondence between the extreme points of the set $\mathcal{S}_{\text{sym}}(\overset{n}{\otimes}\mathcal{H})$ and the atoms of the lattice of symmetric orthogonal projections that is given by the restriction of the support function to the extreme elements.

(iii) Any $W \in \mathcal{S}(\overset{n}{\otimes}\mathcal{H})$ admits a barycentric integral representation in terms of extreme elements

$$W = \int d\mu_W(P)\,\frac{P}{\operatorname{tr}(P)} \tag{2.60}$$

where $\mu_W \in M^1_+(\mathcal{S}(\mathcal{H}))$ is supported by the atoms of $\mathcal{P}_{\text{sym}}(\overset{n}{\otimes}\mathcal{H})$. The σ–field used here is explained in the context of de Finetti's theorem below. The fact that $\mathcal{S}_{\text{sym}}(\overset{n}{\otimes}\mathcal{H})$ is no simplex is equivalent to the fact that the decomposition of a reducible invariant subspace into irreducible components is not unique.

Problem 2.2.1. (i) Are the extreme points of $\mathcal{S}(\mathcal{P}_{\text{sym}}(\overset{n}{\otimes}\mathcal{H}))$ the atoms of $\mathcal{P}_{\text{sym}}(\overset{n}{\otimes}\mathcal{H})$? (ii) Can we identify $\mathcal{S}_{\text{sym}}(\overset{n}{\otimes}\mathcal{H})$ and $\mathcal{S}(\mathcal{P}_{\text{sym}}(\overset{n}{\otimes}\mathcal{H}))$?

2.2.2 Parastatistics

Pure BE symmetric or pure FD symmetric states are distinguished extreme points of $\mathcal{S}_{\text{sym}}(\overset{n}{\otimes}\mathcal{H})$ insofar as they extreme points of $\mathcal{S}(\overset{n}{\otimes}\mathcal{H})$ too.

Definition 2.2.1. $W \in \operatorname{ex}(\mathcal{S}_{\text{sym}}(\overset{n}{\otimes}\mathcal{H}))$ *is called a* normal statistics *if* $W \in \operatorname{ex}(\mathcal{S}(\overset{n}{\otimes}\mathcal{H}))$, *otherwise it is called a* parastatistics.

Remark 2.2.2. For a classification of parastatistics we refer to the work by HARTLE, STOLT and TAYLOR [138, 85] and – in the framework of local quantum field theory – to the monograph by HAAG [84].

Lemma 2.2.7. *MB symmetric statistical operators are no parastatistics.*

Proof. If W is MB symmetric and pure then W is BE symmetric and therefore no parastatistics. Assume that W is MB symmetric and not pure, then $W = \overset{n}{\otimes} w$ where $w \in \mathcal{S}(\mathcal{H})$ is not pure. Moreover, $\operatorname{ran}(W) = \overset{n}{\otimes} \operatorname{ran}(w)$ where $\dim(\operatorname{ran}(w)) > 1$ such that $\operatorname{ran}(W)$ is the tensor product of subspaces with dimension strictly larger than one. Therefore $\operatorname{ran}(W)$ is reducible such that W is no parastatistics. QED

Precisely the parastatistics or convex combinations built up from them according to the integral representation (2.60) never have been observed. To take into account this fact and the empirical nonexistence of convex combinations of BE symmetric and FD symmetric states we formulate the following version of the symmetrization postulate.

Symmetrization Postulate:

- (i a) Linear combinations of FD symmetric state vectors and BE symmetric state vectors do not occur, i.e. Π_+ and Π_- are superselection operators.
- (i b) Mixtures of FD symmetric and BE symmetric states do not occur.
- (ii) Parastatistics do not occur, i.e. only rank–one projections P are admitted in the integral representation (2.60). In other words, the representing measure μ_W is supported by the symmetric orthogonal projections corresponding to one–dimensional representations of the symmetric group.

Notice that by marginalization and the decomposition property (2.20) of the Hilbert space of two particles postulate (ii) is implied by postulate (i b).

The symmetrization postulate implies the nonexistence of mixed MB symmetric states. Nonetheless, these states emerge in the theory of the weak-coupling limit of the BCS model [66]. On the other hand, PARTHASARATHY [125] has drawn attention to certain functional properties of the of the symmetric and antisymmetric Fock spaces that do not exist for the free Fock space.

For a theoretical foundation of the exclusion of parastatistics one usually refers to the 'spin–statistics theorem' (see e.g. [140]) of relativistic quantum field theory. Other arguments are based upon the existence of a complete set of commuting symmetric observables [74, 95] or on the homotopy group of the configuration space (2.30) where $d = 3$ [105]. We shall motivate the exclusion of parastatistics from the viewpoint of extendibility below.

2.2.3 Extendibility

The classification of indistinguishable particles provided by extremality properties is not very enlighting, in particular because of the empirical nonexistence of most extreme points. Here we propose another approach to the classification of symmetric states, namely the concept of extendibility which allows for a classification inside the two familiar quantum statistics.

This concept is adopted from the theory of interchangeable random variables (see Section 3.1). Its physical meaning can be easily explained from the properties of classical FD statistics when n particles are distributed onto d cells. It is well known that it is not possible to enlarge such a system arbitrarily (to add identical particles without destroying the symmetry type of the state). This is, in principle, due to the statistical correlations of the particles which are compatible with a certain maximal number of particles only.

To analyze this phenomenon we introduce some notation. With any $b \in \mathcal{B}(\mathcal{H})$ we associate the extensive observable $S_n(b) \in \mathcal{B}_{\text{sym}}(\overset{n}{\otimes}\mathcal{H})$

$$S_n(b) = (n-1)!\, E_{\text{sym}}(b \otimes \mathbf{1} \otimes \cdots \otimes \mathbf{1}). \tag{2.61}$$

Moreover, generic elements of $\mathcal{S}(\mathcal{B}(\overset{n}{\otimes}\mathcal{H}))$ are denoted by E_n or simply by E if the number of objects is not relevant. If E_n is a symmetric state on $\overset{n}{\otimes}\mathcal{B}(\mathcal{H})$ (for the definition see [93]) we denote by $E_m, 1 \leq m < n$, the marginal states.

Lemma 2.2.8. *Assume that E_n is a symmetric state on $\overset{n}{\otimes}\mathcal{B}(\mathcal{H}), n \geq 2$, then for any $b \in \mathcal{B}_{sa}(\mathcal{H})$ the inequality*

$$\{E_1(b^2) - E_1^2(b)\} + (n-1)\{E_2(b \otimes b) - E_1^2(b)\} \geq 0 \tag{2.62}$$

holds.

Proof. We start from the observation that

$$E_n(\{S_n(b) - E_n(S_n(b))\}^+ \{S_n(b) - E_n(S_n(b))\}) \geq 0, \tag{2.63}$$

which, because of the self-adjointness of b, is equivalent to

$$E_n S_n^2(b) - \{E_n S_n(b)\}^2 \geq 0, \tag{2.64}$$

from which the assertion follows from the symmetry of E_n irrespective of the commutation relations that are assumed between the different factors of the tensor product $\overset{n}{\otimes}\mathcal{B}(\mathcal{H})$. QED

Assume that $b \in \mathcal{B}_{sa}(\mathcal{H})$, that $E_2 \in \mathcal{S}(\mathcal{B}(\overset{2}{\otimes}\mathcal{H}))$ and that the variance

$$\text{Var}(b) = E_1(b^2) - E_1^2(b) \tag{2.65}$$

is strictly positive, then we define the auto–correlation of b by

$$\text{Cor}(b) = \frac{E_2(b \otimes b) - E_1^2(b)}{E_1(b^2) - E_1^2(b)}. \tag{2.66}$$

Notice that $|\text{Cor}(b)| \leq 1$.

Lemma 2.2.9. Correlation Lemma. *Assume that E_n is a symmetric state on $\overset{n}{\otimes}\mathcal{B}(\mathcal{H}), n \geq 2$, and that*

$$\min\{\mathit{Cor}(b); b \in \mathcal{B}_{sa}(\mathcal{H}),\ \mathit{Var}(b) > 0\} = -\frac{1}{r} \tag{2.67}$$

is strictly negative, i.e. $r \geq 1$, then $n \leq r+1$.

Proof. This follows immediately from the preceding lemma. QED

Remark 2.2.3.

(a) Assume that n is arbitrarily large, then

$$\min\{\mathrm{Cor}(b); b \in \mathcal{B}_{sa}(\mathcal{H}), \mathrm{Var}(b) > 0\} \geq 0. \tag{2.68}$$

(b) The number n of particles compatible with a state satisfying inequality (2.67) is necessarily finite, $n \leq [r+1]$, i.e. the correlation structure of a system of indistinguishable particles determines the maximal number of particles compatible with that structure.

Definition 2.2.2. *A statistical operator $W_n \in \mathcal{S}_{\mathrm{sym}}(\overset{n}{\otimes} \mathcal{H})$ of a certain symmetry type (symmetric, MB symmetric, BE symmetric, FD symmetric) is called*

1. N–extendible, *where $N \geq n, N \in \mathbf{N}$, if there exists a statistical operator $W_N \in \mathcal{S}_{\mathrm{sym}}(\overset{N}{\otimes} \mathcal{H})$ of the same symmetry type (symmetric, MB symmetric, BE symmetric, FD symmetric) such that W_n is the marginal state of W_N,*
2. max(N)–extendible, *if W_n is N-extendible but not $(N+1)$-extendible,*
3. ∞–extendible *if it is N-extendible for any $N \in \mathbf{N}$,*
4. unextendible *if it is not ∞-extendible.*

Remarks.

1. An extension, if it exists at all, is not necessarily unique.
2. There are no extensions to different statistics.
3. MB symmetric states are ∞–extendible.
4. FD symmetric states are unextendible (see Lemma 2.2.11 below).
5. There are two types of bosons, namely ∞–extendible bosons and unextendible bosons. In quantum optics these two types are related to classical states (i.e. states with classical structural analogies) and nonclassical states (states exhibiting typical quantum features, e.g. antibunching).
6. Parastatistics on $\overset{n}{\otimes} \mathcal{H}$ are max(n)–extendible. The proof follows immediately.

Non–Extendibility of Parastatistics.

Lemma 2.2.10. *Assume that $W_n \in \mathrm{ex}(\mathcal{S}_{\mathrm{sym}}(\overset{n}{\otimes} \mathcal{H}))$ is a parastatistics, then (i) the marginal statistical operator W_{n-1} is no parastatistics in $\mathcal{S}_{\mathrm{sym}}$ $(\overset{n-1}{\otimes} \mathcal{H})$ and (ii) W_n is not one step extendible.*

Proof. Assertion (i) follows from a classical result by WEYL [153]. For (ii) we argue as follows. Assume that $W_{n+1} \in \mathcal{S}_{\mathrm{sym}}(\overset{n+1}{\otimes} \mathcal{H})$, then the reduced state W_n is not a parastatistics in $\mathcal{S}_{\mathrm{sym}}(\overset{n}{\otimes} \mathcal{H})$ which, in turn, follows from the fact that W_n is not a parastatistics in $\mathcal{S}_{\mathrm{sym}}(\overset{n}{\otimes} \mathcal{H})$ if $W_{n+1} \in \mathrm{ex}(\mathcal{S}_{\mathrm{sym}}(\overset{n+1}{\otimes} \mathcal{H}))$. The latter statement follows from the above mentioned result by WEYL and from the invariance of BE statistics and FD statistics under marginalization. QED

It follows that parastatistics are isolated systems in the sense of invariance of extremality properties under addition/subtraction of particles.

2.2.4 The Theorem of de Finetti

The set of ∞–extendible statistical operators can be characterized explicitly. This is one aspect of the quantum generalization of de Finetti's classical theorem. There are three versions of the theorem

1. an abstract C*–variant by STØRMER [139], the proof being done by Choquet theory,
2. the version considered here, due to HUDSON and MOODY [93], that is appropriate for physical purposes, and
3. a variant by ACCARDI and LU [3], generalizing the previous results to a continuous setting.

The probability measures we are going to establish are constructed by means of Choquet theory and require, therefore, a compact convex set where the σ–field is defined. The topological dual space of $\mathcal{B}(\overset{n}{\otimes}\mathcal{H})$ is equipped to this end with the weak* topology where the set $\mathcal{S}(\mathcal{B}(\mathcal{H}))$ is weak*–compact. The set of statistical operators can be identified with the weak*–dense convex subset of normal states in $\mathcal{S}(\mathcal{B}(\mathcal{H}))$. This set is not weak*–compact (it is, however, norm–closed in the Banach space of trace class operators) but it is measurable (see [93]) and the set of probability measures $M_+^1(\mathcal{S}(\mathcal{H}))$ is defined on the σ–field generated by the relatively open subsets of the normal states.

Theorem.

Theorem 2.2.3. *[93] Assume that* $W_m \in \mathcal{S}_{\text{sym}}(\overset{m}{\otimes}\mathcal{H})$ *is* ∞*–extendible, then there exists an uniquely determined probability measure* $\nu \in M_+^1(\mathcal{S}(\mathcal{H}))$ *such that for any marginal statistical operator* W_n *of the extension an integral representation in terms of MB symmetric states exists*

$$W_n = \int d\nu(w) \overset{n}{\otimes} w, \tag{2.69}$$

where ν *is supported by the pure elements of* $\mathcal{S}(\mathcal{H})$ *(i.e.* $\nu(\text{ex}(\mathcal{S}(\mathcal{H}))) = 1$*) iff any* W_n *is BE symmetric.*

Remark 2.2.4.

(a) The proof, based upon CHOQUET's theory of integral representation of points of elements of convex compact sets in topological vector spaces, reveals that ν is the representing measure of the barycentric decomposition of the extended state on $\overset{\text{N}}{\otimes}\mathcal{B}(\mathcal{H})$ into extreme symmetric elements.

(b) In the classical framework the representing measure is determined by the LLN for the underlying interchangeable random variables. In the quantum case this connection still holds but is connected with the difficulty to find a suitable parametrization for the manifolds $\mathcal{S}(\mathcal{H})$ or $\mathrm{ex}(\mathcal{S}(\mathcal{H}))$. For the simplest case, $\mathcal{H} = \mathbf{C}^2$, we refer to [21].
(c) For simple examples we refer to Section 2.3 and [24].
(d) For the convergence of a sequence of ∞–extendible symmetric states to the classical states of the harmonic oscillator [43, 23]

$$W = \int d\mu(z) < \Psi(z), \cdot > \Psi(z), \tag{2.70}$$

where $\mu \in M_+^1(\mathbf{C})$ and $\Psi(z), z \in \mathbf{C}$ is a coherent state vector ($\Phi(k), k \in \mathbf{Z}_+$ are the number state vectors)

$$\Psi(z) = \exp(-\frac{|z|^2}{2}) \sum_{n=0}^{\infty} \frac{1}{\sqrt{n!}} z^n \Phi(n), \tag{2.71}$$

we refer to [24, 21]. For an explanation of the technical details to apply a generalization of Choquet theory to the non compact set $\mathbf{C}$ see [23].
(e) It is well known (cf. eq.(2.60)) that the barycentric representation of a state in terms of pure states is not unique because $\mathcal{S}(\mathcal{B}(\mathcal{H}))$ is no simplex. HUDSON [92] has drawn attention to the fact that for an ∞–extendible state there is one distinguished representing measure, namely that one determined by de Finetti's theorem. The non uniqueness of a representing measure for $E_1 \in \mathcal{S}(\mathcal{B}(\mathcal{H}))$ can be explained by the fact that only a small amount of the information stored in the unique representing measure for $E \in \mathcal{S}(\overset{\mathbf{N}}{\otimes} \mathcal{B}(\mathcal{H}))$ is necessary for the one–particle–subsystem described by E_1.

Lemma 2.2.11. *[17] FD symmetric statistical operators are unextendible.*

Proof. Assume that $W_n \in \mathcal{S}_{\mathrm{sym}}(\overset{n}{\otimes} \mathcal{H})$ is FD symmetric and ∞–extendible. Then de Finetti's theorem applies and we have in particular for the marginal state W_2

$$W_2 = \int d\nu(w)\, w \otimes w. \tag{2.72}$$

Since W_n is FD symmetric W_2 is FD symmetric such that for the transposition U_τ that exchanges particle 1 and 2 the property $U_\tau W_2 = -W_2$ holds. Using the integral representation (2.72) and taking the trace yields – if we take into account eq.(2.43) – a contradiction

$$-1 = -\mathrm{tr}(W_2) = \int d\nu(w)\, \mathrm{tr}\{U_\tau\, w \otimes w\} = \int d\nu(w)\, \mathrm{tr}(w^2) > 0. \tag{2.73}$$

QED

Remark 2.2.5.
(a) Another argument to demonstrate the unextendibility of FD symmetric states is due to COLEMAN [41] who also considers the (unsolved) problem (called N–representability) to determine how far a given FD symmetric state is extendible (cf. [42]).
If, for negative correlations, the minimum $-1/r, r \geq 1$, of the auto-correlations can be evaluated explicitly, the correlation Lemma 2.2.9 gives an upper bound on the maximal number $n \leq [r+1]$ of particles compatible with the state. In general, however, such a state is not even max$([r+1])$–extendible (cf. Chapter 3.1).
(b) An analogous argument immediately shows that for an ∞–extendible BE symmetric state the representing measure is supported by the pure normal states.

Correlation Inequalities. By means of de Finetti's theorem quantum states are expressed in terms of classical probability measures such that any type of classical inequalities can be applied. The first correlation inequality in the context of symmetric states we encountered was the correlation Lemma 2.2.9. Other types of correlation inequalities are founded upon correlation inequalities of classical probability theory. Here we give one example, related to Bell's inequality, and refer to another one in Section 2.3.

Lemma 2.2.12. *[17] Assume that E is a locally normal symmetric state on $\bigotimes^{\mathbf{N}} \mathcal{B}(\mathcal{H})$ and suppose that $a, b, c, d \in \mathcal{B}_{sa}(\mathcal{H})$ satisfy $||a||, ||b||, ||c||, ||d|| \leq 1$, then (i)*

$$|E_2(a \otimes b) - E_2(a \otimes c)| \leq 1 - E_2(b \otimes c), \tag{2.74}$$

and (ii)

$$|E_2(a \otimes b) - E_2(a \otimes c)| + E_2(d \otimes b) + E_2(d \otimes c) \leq 2. \tag{2.75}$$

The proof is based on a simple inequality.

Lemma 2.2.13. *Assume that $a, b, c \in [-1, 1]$, then the inequality*

$$|ab - ac| \leq 1 - bc \tag{2.76}$$

holds.

Proof. We start from the obvious inequality $0 \leq (b^2 - 1)(c^2 - 1)$ which is equivalent to $(b - c)^2 \leq (1 - bc)^2$, so that we obtain

$$|b - c| \leq |1 - bc|. \tag{2.77}$$

Since $bc \in [-1, 1]$ the difference $1 - bc$ is nonnegative, such that $|1 - bc| = 1 - bc$. Altogether we obtain

$$|ab - ac| = |a|\,|b - c| \leq |b - c| \leq |1 - bc| = 1 - bc. \tag{2.78}$$

QED

We are able now to perform the proof of Lemma 2.2.12.

Proof. (i) De Finetti's theorem implies

$$\begin{aligned} |E_2(a \otimes b) \quad &- \quad E_2(a \otimes c)| = |\int d\nu(e)\{e(a) \otimes e(b) - e(a) \otimes e(c)\}| \\ &\leq \int d\nu(e)|\{e(a) \otimes e(b) - e(a) \otimes e(c)\}| \end{aligned} \tag{2.79}$$

which, in combination with inequality (2.76), yields

$$\begin{aligned} &\int d\nu(e)|\{e(a) \otimes e(b) - e(a) \otimes e(c)\}| \\ &\leq \int d\nu(e)\{1 - e(b) \otimes e(c)\} = \{1 - E_2(b \otimes c)\}. \end{aligned} \tag{2.80}$$

(ii) Combination of eq.(2.74) with the same inequality where a is replaced by d and c by $-c$ yields

$$\begin{aligned} &|E_2(a \otimes b) - E_2(a \otimes c)| + E_2(d \otimes b) + E_2(d \otimes c) \\ \leq \; &|E_2(a \otimes b) - E_2(a \otimes c)| + |E_2(d \otimes b) + E_2(d \otimes c)| \leq 2. \end{aligned} \tag{2.81}$$

QED

Classification. The classification of indistinguishable particles by extremality properties (parastatistics) is, as we have seen, not satisfactory. In combination with the symmetrization postulate we obtain BE statistics and FD statistics but nothing more and – more important – no tool for an inner classification of these states.

We conclude this section with the proposal [17] to classify indistinguishable particles by their extendibility properties. This gives the following three classes of indistinguishable particles (more precisely: of symmetric states)

1. MB statistics – or independent indistinguishable particles. There are two subclasses.
   a) Pure MB symmetric states. These states are BE symmetric too.
   b) Mixed MB symmetric states.
2. ∞–extendible particles (including MB statistics). There are two subclasses.
   a) ∞–extendible bosons.
   b) Mixtures of homogeneous product states of mixed states.
3. Unextendible particles. To this class belong all other species of indistinguishable particles, in particular
   a) unextendible bosons,
   b) fermions,
   c) paraparticles.

Independent Indistinguishable Particles. Whereas in classical probability theory statistical independence (and conditional independence) is the fundament of the theory, by contrast, in quantum probability, – under the assumption of the symmetrization postulate – the concept of independent indistinguishable particles is severely restricted. It is confined to homogeneous product states of pure states which constitute precisely the MB symmetric states of boson systems. In particular, the straightforward quantum structural analogue of a classical (MB symmetric) thermal equilibrium state, a homogeneous product state of one–particle thermal equilibrium states, is not admitted. The most important states of independent indistinguishable particles are the coherent states of the harmonic oscillator.

2.3 Discrete Symmetric Statistics

In this section we introduce elementary examples of discrete symmetric states. In particular, we determine those pure states that are both MB and BE symmetric (coherent states). Moreover, to motivate our definition of the classical three statistics in Section 3.2, we derive these probability distributions here from uniform quantum statistics in an abelian subalgebra of observables. To this end we consider the uniform BE/FD symmetric states as the conditional uniform MB symmetric states. Finally, we stress that many important multiparticle states are symmetric (or limits of symmetric states), e.g. number states, coherent states and (grand) canonical equilibrium states of identical particles.

2.3.1 Quantum Configurations

In this section we confine ourselves to a discrete setting defined by $H = \overset{n}{\otimes}\, \mathbf{C}^d$ where $d, n > 1$. In the Hilbert space $\mathbf{C}^d$ we fix a c.o.n.s. $e_1, \ldots, e_d$ and define one–particle–states by $Q_i = < e_i, \cdot > e_i$.

Definition 2.3.1. *By a* configuration *we understand a vector* $\mathbf{j} \in \{1, \ldots, d\}^n$ *that assigns to each particle a one–particle–state. A* quantum configuration $Q_{\mathbf{j}}$ *is the event corresponding to the configuration* $\mathbf{j} = \{j_1, \ldots, j_n\}$

$$Q_{\mathbf{j}} = Q_{j_1} \otimes \cdots \otimes Q_{j_n}. \tag{2.82}$$

Obviously $Q_{\mathbf{j}} \in \mathcal{P}(\overset{n}{\otimes}\, \mathbf{C}^d), \mathrm{tr}(Q_{\mathbf{j}}) = 1$. Moreover, the set of the d^n quantum configurations corresponds to the orthogonal projections defined by the complete orthonormal system of vectors of $\overset{n}{\otimes}\, \mathbf{C}^d$. Therefore we have $Q_{\mathbf{j}}\, Q_{\mathbf{j}'} = \delta_{\mathbf{j},\mathbf{j}'}\, Q_{\mathbf{j}}$ such that the d^n configurations $Q_{\mathbf{j}}$ generate an abelian subalgebra of observables. In this section we are exclusively concerned with this abelian subalgebra.

Assume that $\overset{n}{\times}\, (\Omega, F)$ is the $n$fold product space of a measurable space (Ω, F). With any $\pi \in S_n$ we associate the bimeasurable bijection $\tilde{\pi} : \overset{n}{\times}\, \Omega \to \overset{n}{\times}\, \Omega$ defined by

$$\tilde{\pi}(\omega_1, \ldots, \omega_n) = (\omega_{\pi(1)}, \ldots, \omega_{\pi(n)}). \tag{2.83}$$

Definition 2.3.2. *A probability measure* μ *defined on the product space* $\overset{n}{\times}\, (\Omega, F)$ *is called* symmetric *if*

$$\tilde{\pi}(\mu) = \mu \tag{2.84}$$

holds for all $\pi \in S_n$.

Definition 2.3.3. *Random variables* $X_1, \ldots, X_n : (\Omega, F, P) \to \mathbf{R}$ *are called* interchangeable *if the induced probability measure* $P^{(X_1, \ldots, X_n)} \in M_+^1(\mathbf{R}^n)$ *is symmetric.*

Lemma 2.3.1. *Assume that* $W \in S(\overset{n}{\otimes} \mathcal{H})$ *and suppose that the probabilities of the random variables* $J_i : (\Omega, F, P) \to \{1, \ldots, d\}, 1 \leq i \leq n)$ *are defined by*

$$P(\mathbf{J} = \mathbf{j}) = \operatorname{tr}(W Q_{\mathbf{j}}), \tag{2.85}$$

then the random variables $J_i, 1 \leq i \leq n$, *are interchangeable iff* W *is symmetric.*

Proof. It is obvious from the definitions that the induced measure $P^{(J_1,\ldots,J_n)}$ is symmetric iff P is symmetric. QED

Definition 2.3.4. *By an* occupation number *we understand a vector* $\mathbf{k} \in \{0, 1, \ldots, n\}^d$ *that assigns to any one-particle-state the number of particles in that state.*

This implies that $\mathbf{k}$ is subject to the constraint $\sum k_i = n$. There are $\binom{d+n-1}{n}$ different occupation numbers. If a configuration $\mathbf{j}$ is given, the associated occupation number $\underline{\kappa}(\mathbf{j})$ is determined by

$$\kappa_i(\mathbf{j}) = \sum_{m=1}^{n} \delta_{j_m, i}, \tag{2.86}$$

which shows that the occupation numbers are invariant under any permutation of the particles, i. e. $\underline{\kappa}(\mathbf{j}) = \underline{\kappa}(\tilde{\pi}(\mathbf{j}))$. For a given occupation number $\mathbf{k}$ the set $\underline{\kappa}^{-1}(\mathbf{k})$ contains precisely $\binom{n}{k_1 \ldots k_d}$ configurations.

2.3.2 Mixed States: Nonuniform and Uniform Statistics

Nonuniform Statistics. Let there be given a statistical operator w on $\mathbf{C}^d$ with spectral representation $w = \sum_m p_m Q_m$. In terms of this one-particle-state we define the *nonuniform MB symmetric statistical operator* on $\overset{n}{\otimes} \mathbf{C}^d$ by

$$W_{MB} = \overset{n}{\otimes} (\sum_m p_m Q_m). \tag{2.87}$$

Lemma 2.3.2. *In the nonuniform MB symmetric state the configuration random variables are independent and identically distributed*

$$P_{MB}(\mathbf{J} = \mathbf{j}) = p_{j_1} \cdots p_{j_n} = \prod_{i=1}^{n} P_{MB}(J_i = j_i). \tag{2.88}$$

Proof. We evaluate the configurations

$$P_{MB}(\mathbf{J} = \mathbf{j}) = \mathrm{tr}(W_{MB}\, Q_{j_1} \otimes \cdots \otimes Q_{j_n}) \tag{2.89}$$

$$= \sum_{i_1,\dots,i_n} < e_{i_1} \otimes \cdots \otimes e_{i_n},\, W_{MB}\, Q_{j_1} \otimes \cdots \otimes Q_{j_n},\, e_{i_1} \otimes \cdots \otimes e_{i_n} >$$

$$= < e_{j_1} \otimes \cdots \otimes e_{j_n},\, \overset{n}{\otimes} (\sum_m p_m Q_m)\, e_{j_1} \otimes \cdots \otimes e_{j_n} > = p_{j_1} \cdots p_{j_n}.$$

QED

In contrast to [125] we do not define W_{BE}, W_{FD} as the normalized restriction of W_{MB} to the BE/FD symmetric subspaces but as conditional statistical operators, i.e. they are defined on $\overset{n}{\otimes}\, \mathbf{C}^d$. Moreover, we consider the statistics on a level of events that is more fundamental than the occupation numbers, namely on the level of the configurations. These observables are, in general, not symmetric.

The *nonuniform BE symmetric statistical operator* is defined as the conditional statistical operator of the nonuniform MB symmetric statistical operator given Π_+

$$W_{BE} = \frac{\Pi_+ W_{MB}\, \Pi_+}{\mathrm{tr}(W_{MB}\, \Pi_+)}. \tag{2.90}$$

Lemma 2.3.3. *For the normalization we obtain a sum over all possible occupation numbers*

$$\mathrm{tr}(W_{MB}\, \Pi_+) = \sum_{\mathbf{k}} p_1^{k_1} \cdots p_d^{k_d}. \tag{2.91}$$

Moreover, the configuration random variables are interchangeable and distributed according to

$$P_{BE}(\mathbf{J} = \mathbf{j}) = \begin{pmatrix} n \\ \kappa_1(\mathbf{j}) \; \dots \; \kappa_d(\mathbf{j}) \end{pmatrix}^{-1} \frac{p_{j_1} \cdots p_{j_n}}{\sum_{\mathbf{k}} p_1^{k_1} \cdots p_d^{k_d}}. \tag{2.92}$$

Proof. We first evaluate the normalization factor

$$\mathrm{tr}(W_{MB}\, \Pi_+) = \sum_{j_1,\dots,j_n} < e_{j_1} \otimes \cdots \otimes e_{j_n},\, W_{MB} \frac{1}{n!} \sum_{\pi} e_{j_{\pi(1)}} \otimes \cdots \otimes e_{j_{\pi(n)}} >$$

$$= \sum_{j_1,\dots,j_n} < e_{j_1} \otimes \cdots \otimes e_{j_n},\, \frac{1}{n!} \sum_{\pi} p_{\pi(1)} \cdots p_{\pi(n)}\, e_{j_{\pi(1)}} \otimes \cdots \otimes e_{j_{\pi(n)}} >$$

$$= \sum_{\mathbf{k}} \sum_{\mathbf{j} \in \underline{\kappa}^{-1}(\mathbf{k})} \frac{1}{n!} \sum_{\pi} p_{\pi(1)} \cdots p_{\pi(n)} \cdot \tag{2.93}$$

$$\cdot < e_{j_1} \otimes \cdots \otimes e_{j_n},\, e_{j_{\pi(1)}} \otimes \cdots \otimes e_{j_{\pi(n)}} >$$

$$= \sum_{\mathbf{k}} \begin{pmatrix} n \\ k_1 \; \dots \; k_d \end{pmatrix} \frac{1}{n!} k_1! \cdots k_d!\, p_1^{k_1} \cdots p_d^{k_d} = \sum_{\mathbf{k}} p_1^{k_1} \cdots p_d^{k_d}.$$

Next we evaluate the nominator for the configurations

$$\begin{aligned}
&\mathrm{tr}(\Pi_+ W_{MB}\, \Pi_+ Q_{j_1} \otimes \cdots \otimes Q_{j_n}) \\
&= \; < \Pi_+ e_{j_1} \otimes \cdots \otimes e_{j_n}, W_{MB}\, \Pi_+ e_{j_1} \otimes \cdots \otimes e_{j_n} > \\
&= \; (\frac{1}{n!})^2 \sum_{\pi',\pi} < e_{j_{\pi'(1)}} \otimes \cdots \otimes e_{j_{\pi'(n)}}, W_{MB}\, e_{j_{\pi(1)}} \otimes \cdots \otimes e_{j_{\pi(n)}} > \\
&= \; \frac{1}{n!} \sum_{\pi} < e_{j_1} \otimes \cdots \otimes e_{j_n}, W_{MB}\, e_{j_{\pi(1)}} \otimes \cdots \otimes e_{j_{\pi(n)}} > \\
&= \; \frac{1}{n!} \sum_{\pi} \prod_{i=1}^{n} < e_{j_i}, e_{j_{\pi(i)}} > p_{j_{\pi(i)}} \; = \; \frac{1}{n!} \mathrm{per}\, \{\mathcal{E}(\mathbf{j})\, \mathcal{D}(\mathbf{p})\}.
\end{aligned} \tag{2.94}$$

Here per$(\cdot)$ denotes the permanent of a square matrix and $\mathcal{E}(\mathbf{j})$ is a real $n \times n$ matrix determined by the configuration $\mathbf{j}$ with entries $\mathcal{E}(\mathbf{j})_{k,\ell} =< e_{j_k}, e_{j_\ell} > \in \{0,1\}$ whereas $\mathcal{D}(\mathbf{p})$ is a real $n \times n$ diagonal matrix determined by the statistical operator W_{MB} and the configuration $\mathbf{j}$ with entries $\mathcal{D}(\mathbf{p})_{k,\ell} = \delta_{k,\ell}\, p_{j_k}$. Using the identity

$$\mathrm{per}\, \{\mathcal{E}(\mathbf{j})\, \mathcal{D}(\mathbf{p})\} \; = \; \mathrm{per}\, \{\mathcal{E}(\mathbf{j})\}\, \mathrm{per}\{\mathcal{D}(\mathbf{p})\} \tag{2.95}$$

we obtain

$$\mathrm{per}\, \{\mathcal{E}(\mathbf{j})\} \; = \; \kappa_1(\mathbf{j})! \cdots \kappa_d(\mathbf{j})!, \quad \mathrm{per}\{\mathcal{D}(\mathbf{p})\} \; = \; p_{j_1} \cdots p_{j_n}, \tag{2.96}$$

from which the assertion follows. QED

For $n \leq d$ the *nonuniform FD symmetric statistical operator* is defined as the conditional statistical operator of the nonuniform MB symmetric statistical operator given Π_-

$$W_{FD} \; = \; \frac{\Pi_- W_{MB}\, \Pi_-}{\mathrm{tr}(W_{MB}\, \Pi_-)}. \tag{2.97}$$

Lemma 2.3.4. *For the normalization we obtain a sum over all possible occupation numbers restricted to* $\{0,1\}^d$

$$\mathrm{tr}(W_{MB}\, \Pi_-) \; = \; \sum_{\mathbf{k} \in \{0,1\}^d} p_1^{k_1} \cdots p_d^{k_d}. \tag{2.98}$$

Moreover, the configuration random variables are distributed according to the symmetric law

$$P_{FD}(\mathbf{J} = \mathbf{j}) \; = \; \begin{cases} (n!)^{-1} \dfrac{p_{j_1} \cdots p_{j_n}}{\sum_{\mathbf{k} \in \{0,1\}^d} p_1^{k_1} \cdots p_d^{k_d}} & \text{if } \; \underline{\kappa}(\mathbf{j}) \in \{0,1\}^d, \\ 0 & \text{else.} \end{cases} \tag{2.99}$$

Proof. The normalization is obtained in analogy to the procedure applied for BE statistics. Using the same steps as in the proof for the distribution of the corresponding quantities in BE statistics we obtain

$$\mathrm{tr}(\Pi_-\, W_{MB}\, \Pi_-\, Q_{j_1} \otimes \cdots \otimes Q_{j_n}) = \frac{1}{n!}\det\{\mathcal{E}(\mathbf{j})\,\mathcal{D}(\mathbf{p})\}, \tag{2.100}$$

where the matrices $\mathcal{E}(\mathbf{j})$ and $\mathcal{D}(\mathbf{p})$ have been introduced in Lemma 2.3.3. For the determinant of $\mathcal{E}(\mathbf{j})$ we obtain

$$\det(\mathcal{E}(\mathbf{j})) = \begin{cases} 1, & \text{for } \underline{\kappa}(\mathbf{j}) \in \{0,1\}^d, \\ 0, & \text{else}, \end{cases} \tag{2.101}$$

since for $\underline{\kappa}(\mathbf{j}) \notin \{0,1\}^d$ there exist at least two identical rows (which are linearly dependent) whereas for $\underline{\kappa}(\mathbf{j}) \in \{0,1\}^d$ there exists a real orthogonal $n \times n$ matrix $\mathcal{O}$ such that $\mathcal{O}\,\mathcal{E}(\mathbf{j})\,\mathcal{O}^+ = \mathbf{1}$ holds. QED

Problem 2.3.1. Can we consider BE/FD statistics as conditioned MB statistics, i.e. do there exist events Y, Z such that

$$P_{BE}(\mathbf{J} = \mathbf{j}) = P_{MB}(\mathbf{J} = \mathbf{j}\,|\,Y), \tag{2.102}$$

$$P_{FD}(\mathbf{J} = \mathbf{j}) = P_{MB}(\mathbf{J} = \mathbf{j}\,|\,Z)? \tag{2.103}$$

Uniform Statistics.

Definition 2.3.5. *We define* uniform *MB/BE/FD statistics by means of the statistical operators (FD statistics requires $n \leq d$)*

$$W_{MB} = \left(\frac{1}{d}\right)^n \mathbf{1}, \tag{2.104}$$

$$W_{BE} = \frac{\Pi_+ W_{MB} \Pi_+}{\mathrm{tr}(\Pi_+ W_{MB})} = \binom{d+n-1}{n}^{-1} \Pi_+, \tag{2.105}$$

$$W_{FD} = \frac{\Pi_- W_{MB} \Pi_-}{\mathrm{tr}(\Pi_- W_{MB})} = \binom{d}{n}^{-1} \Pi_-. \tag{2.106}$$

Lemma 2.3.5. *In the uniform statistics the configurations are given by the interchangeable random variables*

$$P_{MB}(\mathbf{J} = \mathbf{j}) = \left(\frac{1}{d}\right)^n, \tag{2.107}$$

$$P_{BE}(\mathbf{J} = \mathbf{j}) = \binom{n}{\kappa_1(\mathbf{j}) \ldots \kappa_d(\mathbf{j})}^{-1} \binom{d+n-1}{n}^{-1}, \tag{2.108}$$

$$P_{FD}(\mathbf{J} = \mathbf{j}) = \begin{cases} (n!)^{-1} \binom{d}{n}^{-1} & \text{if } \underline{\kappa}(\mathbf{j}) \in \{0,1\}^d, \\ 0 & \text{else}. \end{cases} \tag{2.109}$$

Proof. This follows from Lemmata 2.3.2, 2.3.3, 2.3.4 specialized to the uniform statistics. QED

Remark 2.3.1. Since eqs.(2.105,2.106) hold for all $d, n \geq 2$, $d \geq n$ respectively, and determine the marginal states by means of the same formulae, the uniform BE symmetric states are ∞–extendible and the uniform FD symmetric states are max(d)–extendible. For the integral representation of W_{BE} see Lemma 2.3.9 below.

2.3.3 Pure States: Number States and Coherent States

Number States. $\overset{n}{\otimes}\, \mathbf{C}^d$ is a Hilbert space of dimension d^n whereas

$$\dim((\overset{n}{\otimes}\, \mathbf{C}^d)_+) = \begin{pmatrix} d+n-1 \\ n \end{pmatrix}, \quad \dim((\overset{n}{\otimes}\, \mathbf{C}^d)_-) = \begin{pmatrix} d \\ n \end{pmatrix}, \tag{2.110}$$

such that for $n > d$ no FD symmetric subspace exists. Moreover, for $d, n \geq 2$

$$d^n \geq \begin{pmatrix} d+n-1 \\ n \end{pmatrix} \geq \begin{pmatrix} d \\ n \end{pmatrix}. \tag{2.111}$$

To any occupation number $\mathbf{k}$ there corresponds a *quantum occupation number* defined by

$$\mathcal{Q}_{\mathbf{k}} = \sum_{\mathbf{j} \in \underline{\kappa}^{-1}(\mathbf{k})} Q_{\mathbf{j}}. \tag{2.112}$$

Obviously $\mathcal{Q}_{\mathbf{k}} \in \mathcal{P}_{\mathrm{sym}}(\overset{n}{\otimes}\, \mathcal{H})$ and $\mathrm{tr}(\mathcal{Q}_{\mathbf{k}}) = \begin{pmatrix} & n & \\ k_1 & \ldots & k_d \end{pmatrix}$, but in general the normalized $\mathcal{Q}_{\mathbf{k}}$ are in general neither MB symmetric nor BE/FD symmetric.

To any occupation number $\mathbf{k}$, however, there corresponds one one-dimensional subspace of the BE symmetric subspace of $\overset{n}{\otimes}\, \mathbf{C}^d$ and to any occupation number $\mathbf{k} \in \{0,1\}^d$ there corresponds one one-dimensional subspace of the FD symmetric subspace of $\overset{n}{\otimes}\, \mathbf{C}^d$ which are defined as follows.

Definition 2.3.6. *Given the fixed basis of* $\mathbf{C}^d$, *in* $(\overset{n}{\otimes}\, \mathbf{C}^d)_+$ *and* $(\overset{n}{\otimes}\, \mathbf{C}^d)_-$ *we introduce a distinguished c.o.n.s.* ($\overset{0}{\otimes}$ *means that this factor is omitted) – the* BE symmetric number state vectors *and the* FD symmetric number state vectors –

$$\phi_n^+(\mathbf{k}) = \begin{pmatrix} & n & \\ k_1 & \ldots & k_d \end{pmatrix}^{1/2} \Pi_+ \overset{k_1}{\otimes} e_1 \cdots \overset{k_d}{\otimes} e_d, \tag{2.113}$$

$$\phi_n^-(\mathbf{k}) = (n!)^{1/2}\, \Pi_- \overset{k_1}{\otimes} e_1 \cdots \overset{k_d}{\otimes} e_d, \tag{2.114}$$

where $\mathbf{k}$ *is any occupation number.*

Remark 2.3.2.
(a) Obviously, $\phi_n^-(\mathbf{k}) = 0$ if $\mathbf{k} \notin \{0,1\}^d$.

(b) In terms of these basis vectors the orthogonal projections onto $(\overset{n}{\otimes}\, \mathbf{C}^d)_\pm$ read

$$\Pi_\pm = \sum_{\mathbf{k}} < \phi_n^\pm(\mathbf{k}), \cdot > \phi_n^\pm(\mathbf{k}). \tag{2.115}$$

(c) For $d = 2$ there exists a one-to-one correspondence between $\Pi_+(\overset{n}{\otimes}\, \mathbf{C}^2)$ and the state space of a spin-$n/2$ system that is given by $\phi_n(k) \leftrightarrow |\frac{n}{2}, \frac{2k-n}{2} >, k = 0, 1, \ldots, n$.

Lemma 2.3.6. *The probability distribution of the configurations in number states is given by the Dirac measures (for FD statistics we assume* $\mathbf{k} \in \{0,1\}^d$*)*

$$P_{\mathbf{k}}^\pm(\mathbf{J} = \mathbf{j}) = < \phi_n^\pm(\mathbf{k}), Q_{\mathbf{j}}\, \phi_n^\pm(\mathbf{k}) > = \delta_{\mathbf{k}, \underline{\kappa}(\mathbf{j})}. \tag{2.116}$$

Proof. We obtain

$$\begin{aligned} < \phi_n^\pm(\mathbf{k}), Q_{\mathbf{j}}\, \phi_n^\pm(\mathbf{k}) > &= | < \phi_n^\pm(\mathbf{k}), e_{j_1} \otimes \cdots \otimes e_{j_n} > |^2 \\ &= \begin{pmatrix} & n & \\ k_1 & \ldots & k_d \end{pmatrix} | < \overset{k_1}{\otimes} e_1 \otimes \cdots \otimes \overset{k_d}{\otimes} e_d, \frac{1}{n!} \sum_\pi e_{j_{\pi(1)}} \otimes \cdots \otimes e_{j_{\pi(n)}} |^2 \\ &= \begin{pmatrix} & n & \\ k_1 & \ldots & k_d \end{pmatrix} \frac{k_1! \cdots k_d!}{n!} \delta_{\mathbf{k}, \underline{\kappa}(\mathbf{j})}. \end{aligned} \tag{2.117}$$

QED

Coherent States. Finally we consider the distinguished class of states that are both MB symmetric and BE symmetric.

Any pure state on $\mathcal{B}(\mathbf{C}^d)$ can be defined in terms of a state vector of the form

$$\psi(\mathbf{z}) = \sum_{i=1}^d z_i e_i, \tag{2.118}$$

where $\mathbf{z} \in \mathbf{C}^d, \sum_i |z_i|^2 = 1$. This is a parametrization of states on $\mathbf{C}^d$ in terms of elements of $\mathbf{C}^d$ itself (or of the dual space). To eliminate any redundancy we assume in what follows that $z_1 = \sqrt{1 - \sum_{i=2}^d |z_i|^2} \in \mathbf{R}$ such that these states are parametrized by $d - 1$ complex numbers of the set

$$D^{(d)} = \{(z_2, \ldots, z_d) \in \mathbf{C}^{d-1}; \sum_{i=2}^d |z_i|^2 \leq 1\}. \tag{2.119}$$

This set is topologically identical to the unit ball in $\mathbf{R}^{2(d-1)}$ such that

$$\mathrm{Vol}(D^{(d)}) = \frac{\pi^{d-1}}{(d-1)!}. \tag{2.120}$$

Definition 2.3.7. *The states on $\mathcal{B}(\overset{n}{\otimes}\, \mathbf{C}^d)$ defined in terms of the state vectors*

$$\psi_n(\mathbf{z}) = \overset{n}{\otimes}\, \psi(\mathbf{z}) \tag{2.121}$$

are called discrete coherent states, *$\psi(\mathbf{z})$ is called an* elementary coherent state *vector.*

Lemma 2.3.7. *The configuration random variables in coherent states are i.i.d.*

$$P_{\mathbf{z}}(\mathbf{J}=\mathbf{j}) = <\psi_n(\mathbf{z}),\, Q_{\mathbf{j}}\,\psi_n(\mathbf{z})> = \prod_{i=1}^{n} |z_{j_i}|^2 = \prod_{i=1}^{n} P_{\mathbf{z}}(J_i = j_i) \tag{2.122}$$

such that the number of particles in coherent states is determined by a multinomial law

$$P_{\mathbf{z}}(\mathbf{K}=\mathbf{k}) = |<\psi_n(\mathbf{z}), \phi_n^+(\mathbf{k})>|^2 = M_{n,\mathbf{p}}(\mathbf{k}), \tag{2.123}$$

where $\mathbf{p} = (|z_1|^2, \ldots, |z_d|^2)$.

Proof. For the configurations we obtain

$$\begin{aligned} <\psi_n(\mathbf{z}),\, Q_{\mathbf{j}}\,\psi_n(\mathbf{z})> &= |<\psi_n(\mathbf{z}),\, e_{j_1}\otimes\cdots\otimes e_{j_n}>|^2 \\ &= \prod_{i=1}^{n} |<\psi(\mathbf{z}), e_{j_i}>|^2 = \prod_{i=1}^{n} |z_{j_i}|^2, \end{aligned} \tag{2.124}$$

and for the occupation numbers we have

$$\begin{aligned} <\psi_n(\mathbf{z}),\, \phi_n^+(\mathbf{k})> &= \begin{pmatrix} n \\ k_1 \ldots k_d \end{pmatrix}^{1/2} <\psi_n(\mathbf{z}),\, \Pi_+ \overset{k_1}{\otimes} e_1 \cdots \overset{k_d}{\otimes} e_d> \\ &= \begin{pmatrix} n \\ k_1 \ldots k_d \end{pmatrix}^{1/2} <\psi_n(\mathbf{z}),\, \overset{k_1}{\otimes} e_1 \cdots \overset{k_d}{\otimes} e_d> \\ &= \begin{pmatrix} n \\ k_1 \ldots k_d \end{pmatrix}^{1/2} z_1^{k_1}\cdots z_d^{k_d}, \end{aligned} \tag{2.125}$$

where we used $\Pi_+\,\psi_n(\mathbf{z}) = \psi_n(\mathbf{z})$. QED

For $d = 2$ the following result is well known in those contexts where discrete coherent states are called *atomic coherent states* [6] or *coherent spin states* [133].

Lemma 2.3.8. *[31] The discrete coherent states admit a continuous resolution of the identity on $\Pi_+(\overset{n}{\otimes}\, \mathbf{C}^d)$, i.e.*

$$\int d\mu_d(\mathbf{z}) \begin{pmatrix} d+n-1 \\ n \end{pmatrix} <\psi_n(\mathbf{z}), \cdot>\, \psi_n(\mathbf{z}) = \Pi_+, \tag{2.126}$$

where μ_d is the uniform probability measure on the set $D^{(d)}$

$$d\mu_d(\mathbf{z}) = \frac{(d-1)!}{\pi^{d-1}} 1_{[\sum_{i=2}^{d} |z_i|^2 \leq 1]}(z_2, \ldots, z_d)\, dz_2 \cdots dz_d, \tag{2.127}$$

and the convention $z_1 = \sqrt{1 - \sum_{i=2}^{d} |z_i|^2}$ is supposed.

Proof. Obviously

$$\int d\mu_d(\mathbf{z}) \binom{d+n-1}{n} < \psi_n(\mathbf{z}), \Xi > \psi_n(\mathbf{z}) = 0 \tag{2.128}$$

for any $\Xi \in (\mathbf{1} - \Pi_+)(\overset{n}{\otimes}\, \mathbf{C}^d)$ such that the operator defined by the integral has its range in $\Pi_+(\overset{n}{\otimes}\, \mathbf{C}^d)$. It is, therefore, sufficient to analyse the action on the c.o.n.s. of number states. Setting $dz_i = dp_i d\alpha_i/2, 2 \leq i \leq d$, and integrating the phases gives

$$\int d\mu_d(\mathbf{z}) \binom{d+n-1}{n} < \phi_n^+(\mathbf{k}'), \psi_n(\mathbf{z}) >< \psi_n(\mathbf{z}), \phi_n^+(\mathbf{k}) >$$
$$= \delta_{\mathbf{k},\mathbf{k}'} \int d\nu_d(\mathbf{p}) \binom{d+n-1}{n} M_{n,\mathbf{p}}(\mathbf{k}), \tag{2.129}$$

where $\nu_d(\mathbf{p})$ is the uniform probability measure on the simplex

$$S_d = \{p_2, \ldots, p_d \in \mathbf{R}_+; \sum_{i=2}^{d} p_i \leq 1\} \tag{2.130}$$

with volume

$$\mathrm{Vol}(S_d) = \frac{\mathrm{Vol}(D^{(d)})}{\pi^{d-1}} = \frac{1}{(d-1)!} \tag{2.131}$$

and $M_{n,\mathbf{p}}(\mathbf{k})$ is the multinomial distribution and $p_1 = 1 - \sum_{i=2}^{d} p_i$ is supposed. Since

$$\mathcal{D}_{n,\mathbf{k}}(\mathbf{p}) = (d-1)! \binom{d+n-1}{n} M_{n,\mathbf{p}}(\mathbf{k}) \tag{2.132}$$

is a probability distribution on S_d the assertion follows. QED

Remark 2.3.3.

(a) The probability distribution for the occupation numbers in coherent states is the multinomial distribution (a discrete probability distribution). On the other hand the same formula determines the probability distribution of a coherent event in number states. This allows us to consider the (renormalized) multinomial distribution $M_{n,\mathbf{p}}(\mathbf{k})$ as a probability *density* with parameters $(n, \mathbf{k})$ and variable $\mathbf{p}$. The resulting distribution, eq.(2.132), is a (multivariate) *Dirichlet distribution*. In Bayesian statistics this distribution is known as the *conjugate distribution* of the multinomial distribution. The conjugate distribution of the binomial distribution, $d = 2$, is a *β–distribution*. In this sense the discrete coherent states are the quantum conjugate states of the BE symmetric number states.

(b) For an infinite dimensional one–particle Hilbert space $\mathcal{H}$ with c.o.n.s. $e_i, i \in \mathbf{N}$ coherent states are parametrized in terms of elements ϕ of $\mathcal{H}' = \mathrm{cl}(\mathrm{lin}\{e_2, e_3, \ldots\})$ where $\phi_1 =< e_1, \phi >= \sqrt{1 - \sum_{i=2}^{\infty} |< e_i, \phi >|^2} \in \mathbf{R}$ is supposed, such that de Finetti's integral representation for bosons, formulated in terms of integration on Hilbert space, reads

$$W_n = \int_{\mathcal{H}'} d\mu(\phi) < \psi_n(\phi), \cdot > \psi_n(\phi), \tag{2.133}$$

where the probability measure is supported by the unit ball of $\mathcal{H}'$ so that $\mu(\{\phi \in \mathcal{H}; ||\phi|| > 1\}) = 0$.

(c) A related integral representation [43], on the full Hilbert space, emerges from the generalization of the classical states of the harmonic oscillator, cf. eq.(2.70), to the classical states of the quantized electromagnetic field (states on the boson Fock space over $\mathcal{H}$)

$$W = \int_{\mathcal{H}} d\mu(\phi) < \Psi(\phi), \cdot > \Psi(\phi), \tag{2.134}$$

where for $\phi \in \mathcal{H}$ the vector $\Psi(\phi)$ denotes a coherent state vector (cf. eq.(2.71) for the (symmetric) Fock space over $\mathbf{C}$)

$$\Psi(\phi) = \exp(-\frac{||\phi||^2}{2}) \bigoplus_{n=0}^{\infty} \{ \frac{1}{\sqrt{n!}} \overset{n}{\otimes} \phi \} \tag{2.135}$$

and μ is any probability measure on $\mathcal{H}$.

Lemma 2.3.9. *[31] The uniform BE symmetric state is determined by the uniform probability measure μ_d on $D^{(d)}$*

$$\int d\mu_d(\mathbf{z}) < \psi_n(\mathbf{z}), \cdot > \psi_n(\mathbf{z}) = \binom{d+n-1}{n}^{-1} \Pi_+. \tag{2.136}$$

Proof. This follows immediately from Lemma 2.3.8. QED

2.3.4 An Inequality for an Occupancy Event

Definition 2.3.8. *By an* occupancy number *we understand a vector $\mathbf{z} \in \{0, 1, \ldots, d\}^{n+1}$ that determines the number of one–particle–states with a given number of particles.*

This implies that $\mathbf{z}$ is subject to the constraints $\sum z_i = d, \sum i z_i = n$. If an occupation number $\mathbf{k}$ is given, the associated occupancy number $\underline{\zeta}(\mathbf{k})$ is determined by

$$\zeta_i(\mathbf{k}) = \sum_{m=1}^{d} \delta_{k_m, i}, \tag{2.137}$$

Here we are concerned with the *quantum occupancy number* $\sum_{i=1}^{d} \overset{n}{\otimes} Q_i$ corresponding to the special occupancy number $\mathbf{z} = (d-1, 0, \ldots, 0, 1)$ where all particles are confined to one cell.

Lemma 2.3.10. *[17] Assume that E is a symmetric state on $\overset{\mathbf{N}}{\otimes} \mathcal{B}(\mathbf{C}^d)$, then for all $n \in N$ the inequality*

$$E_n(\sum_{i=1}^{d} \overset{n}{\otimes} Q_i) \geq \frac{1}{d^{n-1}} \tag{2.138}$$

is satisfied, where equality holds iff the de Finetti measure ν of E is supported by those states $e \in \mathcal{S}(\mathcal{B}(\mathbf{C}^d))$ that satisfy $e(Q_i) = 1/d, 1 \leq i \leq d$.

Proof. De Finetti's theorem yields

$$E_n(\sum_{i=1}^{d} \overset{n}{\otimes} Q_i) = \int d\nu(e) \sum_{i=1}^{d} (\overset{n}{\otimes} e)(\overset{n}{\otimes} Q_i) = \int d\nu(e) \sum_{i=1}^{d} \{e(Q_i)\}^n \tag{2.139}$$

and Hölder's inequality implies

$$\sum_{i=1}^{d} \{e(Q_i)\}^n \geq \frac{1}{d^{n-1}} \tag{2.140}$$

where equality holds iff $e(Q_i) = 1/d, 1 \leq i \leq d$. QED

Remark 2.3.4.
(a) ∞–extendibility implies a certain minimal weight for the probability that all particles are confined to some cell.
(b) MB symmetric states are ∞–extendible so that the inequality is trivially satisfied. The minimum is attained e.g. for the uniform MB symmetric state.
(c) For the nonuniform BE symmetric state W_{BE} we have

$$\mathrm{tr}(W_{BE} \sum_{i=1}^{d} Q_i \otimes \cdots \otimes Q_i) = \frac{\sum_{i=1}^{d} p_i^n}{\sum_{\mathbf{k}} p_1^{k_1} \cdots p_d^{k_d}}. \tag{2.141}$$

such that for the uniform BE symmetric statistical operator W_{BE} we obtain

$$\mathrm{tr}(W_{BE}\sum_{i=1}^{d} Q_i \otimes \cdots \otimes Q_i) = d \begin{pmatrix} d+n-1 \\ n \end{pmatrix}^{-1} \geq \frac{1}{d^{n-1}}, \quad (2.142)$$

where the last inequality follows from eq.(2.111).

(d) An example for a BE symmetric state that violates the inequality is given by eq.(2.55).

(e) For any FD symmetric state the inequality is violated which implies that these states are unextendible.

(f) For the BE symmetric number states we obtain

$$< \phi_n^+(\mathbf{k}), \sum_{i=1}^{d} \{\overset{n}{\otimes} Q_i\} \phi_n^+(\mathbf{k}) > = \sum_{i=1}^{d} \delta_{\mathbf{k},\underline{\kappa}((i,\ldots,i))} = \begin{cases} 1 \text{ if } \mathbf{k} \in \{0,n\}^d, \\ 0 \text{ else}, \end{cases} \quad (2.143)$$

such that those number states which are not MB symmetric are unextendible. They are, for $d = 2$, according to the next lemma, in fact not one step extendible.

Lemma 2.3.11. *Assume $d = 2$. The pure states induced by the vectors $\phi_n^{\pm}(\mathbf{k})$ are not one–step–extendible unless they are MB symmetric (i.e. $\mathbf{k} \in \{0,n\}^d$) in which case they are ∞–extendible.*

Proof. Assume $W_n^{\pm} =< \phi_n(\mathbf{k}), \cdot > \phi_n(\mathbf{k})$ is ∞–extendible, then for all occupation numbers $\mathbf{k} \in \{0,\ldots,n\}^d, \sum_{i=1} k_i = n$, the random variables $J_i : (\Omega, F, P_{\mathbf{k}}) \to \{0,\ldots,d\}, 1 \leq i \leq n$, defined by

$$P_{\mathbf{k}}(\mathbf{J} = \mathbf{j}) = \begin{pmatrix} & n & \\ \kappa_1(\mathbf{j}) & \ldots & \kappa_d(\mathbf{j}) \end{pmatrix}^{-1} \delta_{\mathbf{k},\underline{\kappa}(\mathbf{j})} \quad (2.144)$$

are ∞–extendible. According to Lemma 3.2.1 below, however, for $d = 2$ these random variables are not one step extendible for $\mathbf{k} \in \{0,\ldots,n\}^d, \mathbf{k} \notin \{0,n\}^d$. This implies the thesis. QED

Remarks. Obviously occupation events and occupancy events can be analyzed in the quantum setting. Since, however, these events are contained in the abelian subalgebra generated by the quantum configurations it is more convenient to evaluate the probabilities for them in the classical framework which is introduced in the next chapter.

3. Indistinguishable Classical Particles

In this chapter the concept of indistinguishable classical particles is defined. These are identical classical particles in a state that is characterized by a symmetric probability measure. Since the theory of interchangeable random variables is equivalent to the theory of symmetric probability measures, we prefer to formulate the theory by means of interchangeable random variables. This allows to deal directly with the observables (random variables) and offers several technical advantages.

Whereas the fundamentals for interchangeable random variables are presented in some detail, the application to indistinguishable particles is restricted in this work to a discrete setting (distribution of particles onto cells). Therefore the theory of indistinguishable particles in a continuous setting (configuration space, phase space) and, in particular, scattering phenomena of indistinguishable particles are not considered. On the other hand, limit laws for discrete indistinguishable particles are considered at various levels (macroscopic limit, continuum limit).

Finally, we remark that de Finetti's theorem is not considered in this chapter but in Chapter 4. In what follows, we do not assume that the reader is familiar with the quantum part of the theory (Chapter 2). Therefore some definitions and remarks are repeated here.

3.1 Indistinguishability and Interchangeability

Interchangeable random variables have been introduced systematically by DE FINETTI [44, 45] for the description of indistinguishable but not necessarily independent repetitions of experiments (trials) in the context of DE FINETTI'S subjectivistic interpretation of the probability concept. For the prehistory of this concept and for an analysis of DE FINETTI'S early contributions we refer to [69, 151] and the references given there. DE FINETTI'S original motivation (see [44]) was the analysis of sequences of random events that are characterized by joint probabilities that do not depend upon the order of the events.

As far as the terminology is concerned, *'interchangeable'* is used synonymously with *'exchangeable'*. We prefer 'interchangeable' because of the unisonance with 'indistinguishable'.

The only textbook that treats the theory of interchangeable random variables in some detail is the monograph by CHOW and TEICHER [40]. A review article by ALDOUS [4] collects many results on interchangeable random variables and details are given in the proceedings of the international conference on 'Exchangeability in Probability and Statistics' [103]. Brief introductions are due to GALAMBOS [73, 71] and special topics are treated in [143].

3.1.1 Interchangeable Random Variables

Interchangeability and Indistinguishability. Assume that $\overset{n}{\times}(\Omega, F)$ is the $n$fold product space of a measurable space (Ω, F). With any $\pi \in S_n$ we associate the bimeasurable bijection $\tilde{\pi} : \overset{n}{\times}\, \Omega \to \overset{n}{\times}\, \Omega$ defined by

$$\tilde{\pi}(\omega_1, \ldots, \omega_n) = (\omega_{\pi(1)}, \ldots, \omega_{\pi(n)}). \tag{3.1}$$

Definition 3.1.1. *A probability measure μ defined on the product space $\overset{n}{\times}$ (Ω, F) is called* symmetric *if*

$$\tilde{\pi}(\mu) = \mu \tag{3.2}$$

holds for all $\pi \in S_n$.

Definition 3.1.2. *Particles are called* identical *if they agree in all their intrinsic (i.e. state independent) properties.*

Remark 3.1.1.
(a) The existence of identical particles is the fundamental assumption of all variants of atomism from antiquity until today.
(b) If the particles are described by a Hamiltonian H they are identical iff H is symmetric.

Definition 3.1.3. *A system of identical classical particles is called a system of* indistinguishable particles *if the system is in a state that is characterized by a symmetric probability measure.*

Remark 3.1.2.
(a) *Indistinguishability is a genuine probabilistic concept.* In a classical deterministic setting no meaningful concept of indistinguishability exists since this implies that all particles are confined to the same point in configuration space or phase space which is in conflict with the impenetrability property of the classical particle concept (cf. Section 1.2).
(b) *There are no indistinguishable particles.* Indistinguishability is a property of the state and not of the particles, i.e. indistinguishability is not an ontological property of certain objects but can be prepared or can be avoided by preparations (it may even be time–dependent). For a system of macroscopically many identical particles, however, it is empirically

impossible to prepare a system in a state where the particles of one species are identifiable (e.g. by their positions).
The fact that indistinguishable particles are labeled (by a number) is a property of the description and not an empirical property. Permuation invariance means that the association: particle $\leftrightarrow$ number, is irrelevant. If the kinematics is *a priori* restricted to a state space where only symmetric probability measures are admitted, it is obvious that in this restricted setting identity implies indistinguishability.

(c) *Identity does not imply indistinguishability.* If the state is a function of a symmetric Hamiltonian H, then identity implies indistinguishability. Therefore, identical particles are indistinguishable in the canonical (grand canonical) thermal equilibrium state of an ideal gas.

Definition.

Definition 3.1.4. *Random variables $X_1, \ldots, X_n$ defined on a probability space (Ω, F, P) are called* interchangeable *if*

$$P(X_i < x_i, 1 \leq i \leq n) = P(X_{\pi(i)} < x_i, 1 \leq i \leq n) \tag{3.3}$$

holds for any $\mathbf{x} = (x_1, \ldots, x_n) \in \mathbf{R}^n$ and any $\pi \in S_n$. A countably infinite sequence of random variables is called interchangeable if eq. (3.3) holds locally, that is, for any finite subset.

Remark 3.1.3. Applications of interchangeable random variables to the theory of population genetics are presented in [101]. It has been observed by DE FINETTI [47] that BE statistics and FD statistics can be formulated by means of interchangeable random variables. Due to the fact, however, that MB statistics – describing, according to the traditional viewpoint, distinguishable particles – provides a trivial example for interchangeable random variables, the equivalence of the mathematical concept of interchangeability and the physical concept of indistinguishability has not been recognized (see for example [147]) and DE FINETTI's observation has not been exploited. In general, because physicists maintained for the fundamental three statistics the misleading concept of indistinguishability of early quantum statistics (see Section 5.2), the physical implications of the theory of interchangeable random variables were overlooked. Therefore, one of the most important physical aspects of this theory, the Poisson limit of de Finetti's theorem, is in general not covered by the surveys cited above.

Example 3.1.1. Let there be given random variables $X_i : \Omega \to S, 1 \leq i \leq n$, with a finite state space S and assume that the random variables are uniformly distributed (equal *a priori* probabilities), that is,

$$P(X_i = x_i, 1 \leq i \leq n) = \text{const} \tag{3.4}$$

holds for any $\mathbf{x} \in S^n$, then the random variables $X_1, \ldots, X_n$ are interchangeable.

Example 3.1.2. Assume that the random variables $X_1, \ldots, X_n$ are independent and identically distributed, then these random variables are interchangeable. In fact, by virtue of independence and identical marginals

$$P(X_i < x_i, 1 \leq i \leq n) = \prod_{i=1}^{n} P(X_i < x_i)$$

$$= \prod_{i=1}^{n} P(X_{\pi(i)} < x_i) = P(X_{\pi(i)} < x_i, 1 \leq i \leq n) \tag{3.5}$$

holds for any $\pi \in S_n$ and all $\mathbf{x} \in \mathbf{R}^n$.

The fundamental fact that we can use instead of symmetric probability measures interchangeable random variables is established in the following theorem.

Theorem 3.1.1. *Assume that $P^{\mathbf{X}} \in M_+^1(\mathbf{R}^n)$ is the induced probability measure of a sequence of random variables $\mathbf{X} = (X_1, \ldots, X_n)$ which are defined on a probability space (Ω, F, P), that is, $P^{\mathbf{X}} = P \circ \mathbf{X}^{-1}$. Then the random variables $\mathbf{X}$ are interchangeable if and only if $P^{\mathbf{X}}$ is symmetric.*

Proof. For any $\mathbf{x} \in \mathbf{R}^n$ we set

$$B_{\mathbf{x}} = (-\infty, x_1) \times \cdots \times (-\infty, x_n) \in \mathcal{B}^n,$$

where $\mathcal{B}^n$ denotes the Borel algebra of $\mathbf{R}^n$. Interchangeability is defined by the relation that

$$P(\omega \in \Omega; \mathbf{X}(\omega) \in B_{\mathbf{x}}) = P(\omega \in \Omega; \tilde{\pi}(\mathbf{X}(\omega)) \in B_{\mathbf{x}}) \tag{3.6}$$

holds for any $\pi \in S_n$ and any $\mathbf{x} \in \mathbf{R}^n$. Equivalently,

$$P(\mathbf{X}^{-1}(B_{\mathbf{x}})) = P(\mathbf{X}^{-1}(\tilde{\pi}^{-1}(B_{\mathbf{x}}))) \tag{3.7}$$

or

$$P^{\mathbf{X}}(B_{\mathbf{x}}) = \tilde{\pi}(P^{\mathbf{X}})(B_{\mathbf{x}}) \tag{3.8}$$

holds for any $\pi \in S_n$ and any $\mathbf{x} \in \mathbf{R}^n$. Since the sets $B_{\mathbf{x}}$, $\mathbf{x} \in \mathbf{R}^n$, generate the Borel algebra $\mathcal{B}^n$, eq.(3.8) is equivalent to the symmetry of $P^{\mathbf{X}}$

$$P^{\mathbf{X}} = \tilde{\pi}(P^{\mathbf{X}}). \tag{3.9}$$

QED

Marginals.

Lemma 3.1.1. *(Cf. e.g. [143, proposition 1.1.1.]) Assume that the random variables* $X_1, \ldots, X_n$ *are interchangeable and suppose that for* $m, 1 \leq m < n$, *the indices* $k_1, \ldots, k_m \in \{1, \ldots, n\}$ *are all different from one another, then (i) for any* $\mathbf{x} \in \mathbf{R}^m$ *all m-dimensional marginal probabilities* $P(X_{k_i} < x_i, 1 \leq i \leq m)$ *are identical and in particular identical to* $P(X_i < x_i, 1 \leq i \leq m)$. *(ii) Any subset* $X_{k_1}, \ldots, X_{k_m}$ *of* $X_1, \ldots, X_n$ *is interchangeable.*

Proof. It is sufficient to show that all n marginal probabilities $P(X_1 < x_1, \ldots, X_{m-1} < x_{m-1}, X_{m+1} < x_{m+1}, \ldots, X_n < x_n), 1 \leq m \leq n$, agree and are symmetric. Then we can proceed by induction.

Since

$$\begin{aligned} &P(X_1 < x_1, \ldots, X_{m-1} < x_{m-1}, X_{m+1} < x_{m+1}, \ldots, X_n < x_n) \quad (3.10)\\ &= \quad P(X_1 < x_1, \ldots X_{m-1} < x_{m-1}, X_m < \infty, X_{m+1} < x_{m+1}, \ldots, X_n < x_n) \end{aligned}$$

and this is invariant under permutations, and in particular under the subgroup of permutations where the mth component is fixed, the l.h.s. is invariant under permutations. Using the same equation and permutation invariance shows that all marginal probabilities agree. QED

Remark 3.1.4.

(a) Interchangeable random variables are identically distri"-buted. The converse statement does not hold.
(b) A sequence of interchangeable random variables is stationary. The converse statement does not hold.
(c) Under the appropriate integrability conditions we have

$$E(X_1) = E(X_i), \quad 1 \leq i \leq n, \quad (3.11)$$

$$\mathrm{Var}(X_1) = \mathrm{Var}(X_i), \quad 1 \leq i \leq n, \quad (3.12)$$

$$\mathrm{Cov}(X_1, X_2) = \mathrm{Cov}(X_i, X_j), \quad i \neq j. \quad (3.13)$$

3.1.2 Extendibility

From the properties of independent and identically distributed random variables it is usually considered as self-evident that a sequence of random variables $X_1, \ldots, X_n$ can be extended to a larger sequence $X_1, \ldots X_N$ where $N > n$. If the random variables $X_1, \ldots X_n$ are interchangeable (and describe indistinguishable objects) it is quite natural to require that the sequence $X_1, \ldots X_N$ is itself interchangeable. As we shall see, however, such an embedding into a larger system of indistinguishable objects is not always possible: The correlation structure of the interchangeable sequence $X_1, \ldots, X_n$ can even block the addition of a single object (as it is well known from FD statistics).

Correlation Lemma. There exists a fundamental difference between countably infinite sequences and necessarily finite sequences of interchangeable random variables. One aspect of this difference is connected with the correlation coefficient.

Lemma 3.1.2. Correlation Lemma *(Cf., for the second part, e.g. [4]). Assume that the random variables $X_1, \ldots X_n$ are interchangeable and that $E|X_1|^2 < \infty$ holds.*

1. *Suppose that $Var(X_1) = 0$, then $X_1, \ldots X_n$ are independent.*
2. *Suppose that $0 < Var(X_1)$, then*

$$Cor(X_1, X_2) \geq -\frac{1}{n-1}, \tag{3.14}$$

where equality holds iff $\sum X_i = c$ P-a.e. for some constant c.

Proof. 1. Since $\mathrm{Var}(X_i) = 0$ we have P–a.e. $X_i = E(X_i), 1 \leq i \leq n$. Therefore, for any choice of continuous functions $f_1, \ldots, f_n$, each one vanishing outside a finite interval, we have

$$\begin{aligned} Ef_1(X_1)\cdots f_n(X_n) &= Ef_1(E(X_1))\cdots f_n(E(X_n)) \\ &= f_1(E(X_1))\cdots f_n(E(X_n)) = E(f_1(X_1))\cdots E(f_n(X_n)) \end{aligned} \tag{3.15}$$

such that the random variables $X_1, \ldots X_n$, are independent (cf. [63]).

2. Since the variance of any random variable is nonnegative we have

$$0 \leq \mathrm{Var}(\sum_{i=1}^{n} X_i) = \sum_{i=1}^{n} \mathrm{Var}(X_i) + \sum_{i \neq j} \mathrm{Cov}(X_i, X_j). \tag{3.16}$$

Interchangeability entails that

$$\begin{aligned} \mathrm{Var}(X_i) &= \mathrm{Var}(X_1), 1 \leq i \leq n, & (3.17) \\ \mathrm{Cov}(X_i, X_j) &= \mathrm{Cov}(X_1, X_2), 1 \leq i,j \leq n, i \neq j. & (3.18) \end{aligned}$$

Therefore we obtain

$$0 \leq n\,\mathrm{Var}(X_1) + n(n-1)\,\mathrm{Cov}(X_1, X_2). \tag{3.19}$$

This is equivalent to the inequality

$$0 \leq 1 + (n-1)\,\frac{\mathrm{Cov}(X_1, X_2)}{\mathrm{Var}(X_1)}, \tag{3.20}$$

which may be reformulated as

$$-\frac{1}{n-1} \leq \mathrm{Cor}(X_1, X_2). \tag{3.21}$$

Equality in eq.(3.16) holds iff $\sum_{i=1}^{n} X_i = c$ P-a.e.

QED

Remark 3.1.5.

(a) Assume that the sequence $X_1, \ldots, X_n$ is the segment of a countably infinite sequence, then the correlation coefficient is nonnegative, $\mathrm{Cor}(X_1, X_2) \geq 0$.

(b) Assume that the correlation coefficient of the sequence $X_1, \ldots, X_n$ is strictly negative,

$$\mathrm{Cor}(X_1, X_2) = -\frac{1}{r}, \tag{3.22}$$

where $r \in [1, \infty)$. Eq. (3.14) entails that the number of random variables is necessarily finite

$$n \leq r + 1. \tag{3.23}$$

(c) Strictly positive correlations or a vanishing correlation coefficient do not guarantee that the sequence is ∞–extendible (see the examples below).

(d) An interchangeable sequence $X_1, \ldots, X_n$, with strictly negative correlations
$\mathrm{Cor}(X_1, X_2) = -1/r, r \geq 1$, which is at most ([r+1])-extendible is in general not maximal extendible, i.e. not ([r+1])-extendible. FD statistics is maximal extendible in this sense.

Example 3.1.3. FD Statistics.

Assume that the random variables $X_i : \Omega \to \{1, \ldots, d\}, 1 \leq i \leq n$, are distributed as follows, $\mathbf{j} \in \{1, \ldots, d\}^n$,

$$P(\mathbf{X} = \mathbf{j}) = \begin{cases} \frac{1}{n!} \begin{pmatrix} d \\ n \end{pmatrix}^{-1} & \text{if } \mathbf{k} \in \{0,1\}^d \\ 0 & \text{otherwise,} \end{cases} \tag{3.24}$$

where $\mathbf{k} \in \{0, 1, \ldots, n\}^d$, $k_j = \sum_{m=1}^{n} \delta_{j_m, j}$. This definition makes sense only if $n \leq d$, but it is not *a priori* obvious that this sequence of interchangeable random variables is max(d)–extendible. The correlation coefficient is strictly negative,

$$\mathrm{Cor}(X_1, X_2) = -\frac{1}{d-1}, \tag{3.25}$$

and the correlation lemma implies

$$-\frac{1}{d-1} \geq -\frac{1}{n-1}, \tag{3.26}$$

such that $n \leq d$ follows. This shows that unextendibility is a consequence of specific correlations.

Extendibility.

Definition 3.1.5. *Let there be given a sequence $X_1, \ldots, X_n$ of interchangeable random variables. This sequence is called*

1. N–extendible $N \in \mathbf{N}$, *and $N > n$ if there exists a sequence $Y_1, \ldots, Y_N$ of interchangeable random variables such that*
$$P(X_i < x_i, 1 \le i \le n) = P(Y_i < x_i, 1 \le i \le n) \tag{3.27}$$
holds for any $\mathbf{x} \in \mathbf{R}^n$,
2. max(N)–extendible *if it is N-extendible but not $(N+1)$-extendible,*
3. ∞–extendible *if it is N-extendible for any $N \in \mathbf{N}$,*
4. unextendible *it it is not ∞-extendible.*

Remark 3.1.6.
(a) Since interchangeability is a property of the distribution of random variables and not of the random variables themselves, for an N–extendible sequence $X_1, \ldots, X_n$ the extension is in general denoted by $X_1, \ldots, X_N$.
(b) For an interchangeable sequence $X_1, \ldots, X_n$, DE FINETTI [44, p. 125] established a criterion for checking whether the sequence is ∞–extendible. For more information on the problem of N–extendibilty we refer the reader to [46, 49, 96].
(c) The quantum analogue of the concept of N–extendibility, well–known in the context of FD symmetric states, is called *N–representability* and analyzed for example in [41].

Example 3.1.4. Let there be given interchangeable random variables $X_1, X_2, X_3 : \Omega \to \{0,1\}$ defined by, $p \in [0, 1/3]$, $q = 1 - 3p$,

$$\begin{aligned} P(\mathbf{X} = \epsilon) &= q, && \text{if} \quad \sum \epsilon_i = 0, \\ P(\mathbf{X} = \underline{\epsilon}) &= 0, && \text{if} \quad \sum \epsilon_i = 1, \\ P(\mathbf{X} = \underline{\epsilon}) &= p, && \text{if} \quad \sum \epsilon_i = 2, \\ P(\mathbf{X} = \underline{\epsilon}) &= 0, && \text{if} \quad \sum \epsilon_i = 3, \end{aligned} \tag{3.28}$$

such that these random variables are, according to the lacunae lemma (see Lemma 3.2.1), not 4-extendible if $p > 0$. The following properties hold:

1. These random variables have strictly positive correlation iff $p \in (0, 1/4)$.
2. They are uncorrelated iff $p \in \{0, 1/4\}$. For $p = 1/4$ they are dependent.
3. The random variables are independent iff $p = 0$.
4. For $p = 1/4$ they demonstrate that an extension of the i.i.d random variables $Y_1, Y_2 : \Omega \to \{0,1\}, P(Y_1 = 1) = 1/2$, to three interchangeable random variables is not unique.

5. For $p \in (1/4, 1/3]$ the covariance is strictly negative. In particular, for $p = 1/3$ we have $\mathrm{Cov}(X_1, X_2) = -1/9$ such that these random variables are at most 10-extendible. In fact, they are not 4-extendible.

Proof. For the marginal distributions we obtain, $n = 2$,

$$
\begin{aligned}
P(\mathbf{X} = \underline{\epsilon}) &= q, && \text{if} \quad \sum \epsilon_i = 0, \\
P(\mathbf{X} = \underline{\epsilon}) &= p, && \text{if} \quad \sum \epsilon_i = 1, \\
P(\mathbf{X} = \underline{\epsilon}) &= p, && \text{if} \quad \sum \epsilon_i = 2, \qquad (3.29)
\end{aligned}
$$

and for $n = 1$

$$P(X_1 = 0) = q + p, \qquad P(X_1 = 1) = 2p. \qquad (3.30)$$

Therefore we have

$$\mathrm{Cov}(X_1 X_2) = p(1 - 4p) \qquad (3.31)$$

such that $\mathrm{Cov}(X_1, X_2) > 0$ iff $0 < p < 1/4$, $\mathrm{Cov}(X_1, X_2) = 0$ iff $p \in \{0, 1/4\}$. Moreover,

$$\mathrm{Var}(X_1) = 2p(1 - 2p) \qquad (3.32)$$

such that $\mathrm{Var}(X_1) = 0$ iff $p = 0$ and for this value the random variables are i.i.d. according to the correlation lemma. QED

Classification I. We conclude this section with a proposal [17] to classify indistinguishable classical particles by their extendibility properties. This gives the following three classes of indistinguishable particles (more precisely: of symmetric states)

1. Independent indistinguishable particles. A typical example is MB statistics (see Section 3.2).
2. ∞–extendible particles (including MB statistics). A typical example is BE statistics (see Section 3.2).
3. Unextendible particles. A typical example is FD statistics (see Section 3.2).

Finally we remark that the second class (∞–extendible particles) is characterized by de Finetti's theorem which is considered, in detail, in Chapter 4.

3.2 Levels, Statistics and Groups

In this section we introduce the basic tools for the description of the statistical scheme, namely levels and groups. Moreover, the familiar three statistics are defined.

3.2.1 Levels

Our basic discrete structure is the *statistical scheme* (n, d) which was introduced by BOLTZMANN in 1868 (see Section 5.1). Here n identical particles are distributed onto d identical cells.

Level-1: Configurations. A *configuration* is a vector $\mathbf{j} \in \{1, \ldots, d\}$ where the component $j_i, 1 \leq i \leq n$, denotes that particle i is in cell j_i. If $d = 2$ we usually use another notation, namely $\underline{\epsilon} \in \{0,1\}^n$, for the configurations. There are altogether d^n different configurations. For the probabilistic description we introduce a probability space (Ω, F, P) and random variables.

Definition 3.2.1. *The random variables $J_i : \Omega \to \{1, \ldots, d\}, 1 \leq i \leq n$, where $[J_i = j]$ denotes the event that particle i is in cell $j, 1 \leq j \leq d$, are called* configuration random variables.

It is our basic assumption that the particles are indistinguishable. This assumption is formally expressed by the condition that the configuration random variables are interchangeable, that is,

$$P(J_i = j_i, 1 \leq i \leq n) = P(J_{\pi(i)} = j_i, 1 \leq i \leq n) \tag{3.33}$$

holds for any $\pi \in S_n$ and any $\mathbf{j} \in \{1, \ldots, d\}^n$.

Under this condition some unextendibility properties immediately follow for $d = 2$ if there are isolated lacunae of probability zero (obviously the result is not restricted to $d = 2$).

Lemma 3.2.1. *Assume that the random variables $J_i : (\Omega, F, P) \to \{0,1\}$, $1 \leq i \leq n$, are interchangeable and suppose that for some $k', 0 < k' < n$, the properties $P(\sum_i J_i = k') \neq 0$ and $P(\sum_i J_i = k' \pm 1) = 0$ hold, then the sequence of random variables J_i is not one step extendible.*

Proof. Suppose that the sequence is one step extendible. According to the assumption of interchangeability we have

$$P(\sum_i J_i = k' \pm 1) = \binom{n}{k' \pm 1}^{-1} \sum_{\underline{\epsilon}^{\pm}} P(\mathbf{J} = \underline{\epsilon}^{\pm}) \tag{3.34}$$

$$= \binom{n}{k' \pm 1}^{-1} \{P(\mathbf{J} = \underline{\epsilon}^{\pm}, J_{n+1} = 0) + P(\mathbf{J} = \underline{\epsilon}^{\pm}, J_{n+1} = 1)\}$$

where $\underline{\epsilon}^{\pm} \in \{0,1\}^n$ such that $\sum \epsilon_i = k' \pm 1$. Since, by the assumption of the lemma, this is zero we see that the probabilities $P(\mathbf{J} = \underline{\epsilon}^{\pm}, J_{n+1} = \epsilon_{n+1})$ are zero for $\sum \epsilon_i \in \{k'-1, k', k'+1, k'+2\}$. Therefore,

$$P(\sum^n J_i = k') = \begin{pmatrix} n \\ k' \end{pmatrix}^{-1} \sum_{\underline{\epsilon}, \sum \epsilon_i = k'} P(\mathbf{J} = \underline{\epsilon}) \tag{3.35}$$

$$= \begin{pmatrix} n \\ k' \end{pmatrix}^{-1} \sum_{\underline{\epsilon}, \sum \epsilon_i = k'} \{P(\mathbf{J} = \underline{\epsilon}, J_{n+1} = 0) + P(\mathbf{J} = \underline{\epsilon}, J_{n+1} = 1)\}$$

$$= 0,$$

in contradiction to the assumption. QED

Lemma 3.2.2. *Assume that the random variables* $J_i : (\Omega, F, P_k) \to \{0,1\}$, $1 \le i \le n$, *where* $0 \le k \le n$, *are distributed according to,* $\underline{\epsilon} \in \{0,1\}^n$,

$$P_k(\mathbf{J} = \underline{\epsilon}) = \begin{pmatrix} n \\ k' \end{pmatrix}^{-1} \delta_{k,k'}, \tag{3.36}$$

where $k' = \sum \epsilon_i$ *(classical number states), then, for* $1 \le k \le n-1$ *this sequence of interchangeable random variables is not one step extendible.*

Proof. This follows from Lemma 3.2.1. QED

Level-2: Occupation Numbers. By an *occupation number* we understand a vector $\mathbf{k} \in \{0,1,\ldots,n\}^d$ that assigns to any cell the number of particles in that cell. This implies that $\mathbf{k}$ is subject to the constraint $\sum k_i = n$. There are $\begin{pmatrix} d+n-1 \\ n \end{pmatrix}$ different occupation numbers. If a configuration $\mathbf{j}$ is given, the associated occupation number $\underline{\kappa}(\mathbf{j})$ is determined by

$$\kappa_i(\mathbf{j}) = \sum_{m=1}^{n} \delta_{j_m, i}, \tag{3.37}$$

which shows that the occupation numbers are invariant under any permutation of the particles, i. e. $\underline{\kappa}(\mathbf{j}) = \underline{\kappa}(\tilde{\pi}(\mathbf{j}))$. For a given occupation number $\mathbf{k}$ the set $\underline{\kappa}^{-1}(\mathbf{k})$ contains precisely $\begin{pmatrix} n \\ k_1 \ldots k_d \end{pmatrix}$ configurations.

Since the properties of indistinguishable objects are invariant under permutations, a quantity of particular interest is the number of objects with prescribed properties. To this end we introduce the *indicator random variables*, $1 \le i \le n, 1 \le j \le d$,

$$1_{[J_i = j]}(\omega) = \begin{cases} 1 & \text{if} \quad J_i(\omega) = j, \\ 0 & \text{if} \quad J_i(\omega) \ne j. \end{cases} \tag{3.38}$$

Definition 3.2.2. *If configuration random variables* $J_i : (\Omega, F, P) \to \{1, \ldots, d\}$,
$1 \le i \le n$ *are given, we define the* occupation number random variables $K_j : \Omega \to \{0, 1, \ldots, n\}, 1 \le j \le d$, *by*

$$K_j = \sum_{m=1}^{n} 1_{[J_m = j]}. \tag{3.39}$$

The occupation number random variables are subject to the deterministic constraint

$$\sum_{j=1}^{n} K_j = n \quad \text{P–a.e.} \tag{3.40}$$

Obviously the probabilities $P(\mathbf{K} = \mathbf{k})$ are uniquely determined in terms of the probabilities $P(\mathbf{J} = \mathbf{j})$. Whenever the particles are indistinguishable (interchangeable configurations) the converse statement holds.

Theorem 3.2.1. *[16] Assume that the configuration random variables* $J_i : \Omega \to \{1, \ldots, d\}, 1 \le i \le n$, *are interchangeable then*

1. *for any occupation number* $\mathbf{k}$

$$P(\mathbf{K} = \mathbf{k}) = \binom{n}{k_1 \ldots k_d} P(\mathbf{J} = \mathbf{j}(\mathbf{k})), \tag{3.41}$$

where $\mathbf{j}(\mathbf{k})$ *is any configuration that satisfies* $\underline{\kappa}(\mathbf{j}(\mathbf{k})) = \mathbf{k}$, *and*

2. *for any configuration* $\mathbf{j}$

$$P(\mathbf{J} = \mathbf{j}) = \binom{n}{\kappa_1(\mathbf{j}) \ldots \kappa_d(\mathbf{j})}^{-1} P(\mathbf{K} = \underline{\kappa}(\mathbf{j})). \tag{3.42}$$

Proof. (i) Recall that the set $\underline{\kappa}^{-1}(\mathbf{k})$ contains $\binom{n}{k_1 \ldots k_d}$ configurations that all have, by virtue of interchangeability, the same probability. (ii) We condition upon $[\mathbf{K} = \mathbf{k}]$

$$P(\mathbf{J} = \mathbf{j}) = \sum_{\mathbf{k}} P(\mathbf{J} = \mathbf{j} \,|\, \mathbf{K} = \mathbf{k})\, P(\mathbf{K} = \mathbf{k}). \tag{3.43}$$

Since

$$P(\mathbf{J} = \mathbf{j} \,|\, \mathbf{K} = \mathbf{k}) = \frac{k_1! \cdots k_d!}{n!} \delta_{\underline{\kappa}(\mathbf{j}),\mathbf{k}} \tag{3.44}$$

the result follows. QED

Remark 3.2.1.
(a) Part (i) is used by BOLTZMANN in 1868 and in 1877 (see Section 5.1).
(b) For the generalization of part (ii) to a continuous setting see [119].

Classification II. Theorem 3.2.1 allows us to analyze the classical analogue of the quantum classification according to extreme points considered in Section 2.2.

Lemma 3.2.3. *[17] The set $M^1_{+,\mathrm{sym}}(\{1,\dots,d\}^n)$ of symmetric probability measures on $\{1,\dots,d\}^n$ is a simplex with extreme points (polyhypergeometric distributions)*

$$\mathcal{E}_{\mathbf{k}}(\mathrm{j}) = \begin{pmatrix} & n & \\ \kappa_1(\mathrm{j}) & \dots & \kappa_d(\mathrm{j}) \end{pmatrix}^{-1} \delta_{\mathbf{k},\underline{\kappa}(\mathrm{j})}, \tag{3.45}$$

where $\mathbf{k}$ *is any occupation number.*

Proof. According to Theorem 3.2.1 any symmetric probability distribution on $\{1,\dots,d\}^n$ can be expressed in terms of probability distributions on the set of occupation numbers $\{0,\dots,n\}^d, \sum k_i = n$. This set, however, is a simplex with $\begin{pmatrix} d+n-1 \\ n \end{pmatrix}$ extreme points characterized by the Dirac measures concentrated on the occupation numbers. QED

Lemma 3.2.4. *[17] There are exactly d pure extreme symmetric probability distributions on $\{1,\dots,d\}^n$, characterized by the d cells,*

$$\mathcal{E}_i(\mathrm{j}) = \prod_{m=1}^{n} \delta_{i,j_m}, \tag{3.46}$$

where $1 \leq i \leq d$.

Proof. The set $M^1_+(\{1,\dots,d\}^n)$ of probability measures on $\{1,\dots,d\}^n$ is a simplex with extreme points

$$\mathcal{E}_{\mathrm{j}}(\mathrm{j}') = \delta_{\mathrm{j},\mathrm{j}'}, \tag{3.47}$$

where $\mathbf{j}$ is any occupation number. This agrees with eq.(3.45) iff $j_i = j_1$ for $1 \leq i \leq n$.

QED

Transferring the quantum classification according to extreme points (cf. Section 2.2) to the classical context, we call an extreme symmetric probability distribution a (classical) *normal statistics* if its is pure, otherwise we call it a (classical) *parastatistics*.

The extreme points of $M^1_{+,\mathrm{sym}}(\{1,\dots,d\}^n)$ correspond (up to the normalization) to Dirac measures concentrated on the occupation numbers (they correspond to number states). This is completely different from the quantum framework where extreme elements correspond to invariant irreducible subspaces. Moreover, the quantum normal statistics are pure FD and BE symmetric states. In the classical framework normal statistics are homogeneous product states of pure states (which in the quantum framework are pure MB symmetric states). Compared to the quantum setting this set is rather small. In particular, there are no normal classical statistics with correlations.

Remark 3.2.2. We conclude this analysis of extreme points with a generalization to a continuous setting. The extreme points of the simplex of symmetric probability measures on $\mathbf{R}^n$ are the measures (see e.g. [27])

$$\mathrm{ex}(M^1_{+,\mathrm{sym}}(\mathbf{R}^n)) = \{\frac{1}{n!}\sum_{\pi} \delta_{x_{\pi(1)},\ldots,x_{\pi(n)}} \, ; \, (x_1,\ldots,x_n) \in \mathbf{R}^n\}, \quad (3.48)$$

where $\delta_{\mathbf{x}}$ denotes the Dirac measure concentrated on $\mathbf{x} \in \mathbf{R}^n$. Accordingly, the normal states are given by Dirac measures where all n particles are at the same point. Usually for indistinguishable particles precisely these points are removed from the configuration space (see, e.g. [108]), so that in a continuous setting based upon a configuration space no normal states exist.

Indistinguishable Particles and Indistinguishable Cells. It is obvious that the interchangeability of the configuration random variables $\mathbf{J}$ does not, in general, entail the interchangeability of the associated occupation number random variables $\mathbf{K}$.

Example 3.2.1. The canonical ensemble.

Assume that the particles of the statistical scheme are indistinguishable and that the occupation numbers are distributed according to the law

$$P(\mathbf{K} = \mathbf{k}) = \frac{\exp(-\beta\underline{\varepsilon}\cdot\mathbf{k})}{\sum_{\mathbf{k}}\exp(-\beta\underline{\varepsilon}\cdot\mathbf{k})}, \quad (3.49)$$

where $0 \leq \varepsilon_1 < \ldots < \varepsilon_d$. In this situation the cells are distinguishable by their energies ε_i.

The assumption that the particles and the cells are indistinguishable allows us to infer some more properties.

Lemma 3.2.5. *Assume that the random variables $J_i : \Omega \to \{1,\ldots,d\}, 1 \leq i \leq n$, are interchangeable and that the associated occupation number random variables $K_i : \Omega \to \{0,\ldots,n\}, 1 \leq i \leq d$, are interchangeable, then*

1.

$$Cor(K_1, K_2) = -\frac{1}{d-1}, \quad (3.50)$$

2. and for $1 \leq j \leq d$,

$$E(K_1) = n/d, \quad P(J_1 = j) = 1/d. \quad (3.51)$$

Proof. (1) The constraint (3.40) implies eq.(3.51) according to Lemma 3.1.2. (2) These equations follow immediately from the definition and the interchangeability of the occupation number random variables. QED

Remark 3.2.3. According to eq.(3.50) interchangeable occupation number random variables are max(d)–extendible.

Factorial Moments and Marginal Probabilities. We conclude the analysis of occupation numbers with some results concerning their moments and marginal probabilities. To this end we define the *descending factorials* for $x \in \mathbf{R}$ and $n \in \mathbf{Z}_+$ by

$$x_{[n]} = x(x-1)\cdots(x-n+1), \quad x_{[0]} = 1, \tag{3.52}$$

and the *factorial moments* of a random variable X by $EX_{[k]} = E\{X(X-1)\cdots(X-n+1)\}$.

Lemma 3.2.6. *[13] Assume that the random variables $J_i : \Omega \to \{1,\dots,d\}$, $1 \le i \le n$, are interchangeable, then for any $r, 1 \le r \le n$, and any $k_i, 1 \le k_i \le n, 1 \le i \le r, \sum k_i \le n$, and any $j_i, 1 \le j_i \le d, 1 \le i \le r$ the multidimensional factorial moments of the random variables $K_{j_i}, 1 \le i \le r$, are given by*

$$E\prod_{i=1}^{r}(K_{j_i})_{[k_i]} = \tag{3.53}$$

$$\frac{n!}{(n-\sum k_i)!} P(J_1 = j_1, \dots, J_{k_1} = j_1, J_{k_1+1} = j_2, \dots, J_{k_1+k_2} = j_2, \dots, J_{\sum k_i} = j_r).$$

Remark 3.2.4.
(a) For the related binomial moments we refer to [62, Chapter IV, eq. (5.3)] and [72]. The formula for the factorial moments is due to VON MISES [150] who derives it for the interchangeable occupation number random variables of MB statistics. It holds, however, in general for interchangeable configurations.
(b) For any $j, 1 \le j \le d$, the defining system of $n+1$ linear equations of the factorial moments of one K_j can be solved to express the marginal probabilities $P(K_j = k), 0 \le k \le n$, in terms of the marginal probabilities of $P(J_1 = j, \dots, J_n = j)$. This result of VON MISES [150], however, follows also from the exclusion-inclusion principle which is the basis of the next lemma.

Lemma 3.2.7. *(Cf. e.g. [40, Corrolary 2.1.4.].) Assume that the random variables $J_i : \Omega \to \{1,\dots,d\}, 1 \le i \le n$, are interchangeable, then for any $j, 1 \le j \le d$, and any $k, 0 \le k \le n$ the probabilities of the random variables K_j are given by*

$$P(K_j = k) = \binom{n}{k}\sum_{i=0}^{n-k}(-1)^i\binom{n-k}{i}P(J_1 = j, \dots, J_{k+i} = j). \tag{3.54}$$

Level-3: Occupancy Numbers. By an *occupancy number* we understand a vector $\mathbf{z} \in \{0, 1, \ldots, d\}^{n+1}$ that determines the number of cells with a given number of particles. This implies that $\mathbf{z}$ is subject to the two constraints $\sum z_i = d, \sum i z_i = n$. If an occupation number $\mathbf{k}$ is given, the associated occupancy number $\underline{\zeta}(\mathbf{k})$ is determined by

$$\zeta_i(\mathbf{k}) = \sum_{m=1}^{d} \delta_{k_m, i}. \tag{3.55}$$

Problem 3.2.1. Assume that the statistical scheme (n, d) is given. What is the cardinality of the set of occupancy numbers as a function of n and d?

Definition 3.2.3. *Assume that a sequence of occupation number random variables according to Definition 3.2.2 is given. The random variables* $Z_i : \Omega \to \{0, 1, \ldots, d\}, 0 \leq i \leq n$, *defined by*

$$Z_i = \sum_{m=1}^{d} 1_{[K_m = i]} \tag{3.56}$$

are called occupancy number random variables.

Remark 3.2.5.
(a) The occupancy random variables are subject to two deterministic constraints

$$\sum_{i=0}^{n} Z_i = d, \quad \sum_{i=0}^{n} i Z_i = n, \quad \text{P-a.e.} \tag{3.57}$$

(b) Under the assumption that the occupation number random variables $\mathbf{K}$ are interchangeable we obtain from eq.(3.56) a relation between the expectation of the occupancy number random variables and the probability of the occupation number random variables, $0 \leq k \leq n$,

$$E(Z_k/d) = P(K_1 = k). \tag{3.58}$$

In the following we assume that both the particles and the cells are indistinguishable. Under this condition the combinatorial Theorem 3.2.1 can be reformulated for the occupancy number random variables.

Theorem 3.2.2. *Assume that both the configuration random variables and the occupation number random variables are interchangeable, then*

1. *for any occupation number* $\mathbf{z}$

$$P(\mathbf{Z} = \mathbf{z}) = \binom{d}{z_0 \ldots z_n} P(\mathbf{K} = \mathbf{k}(\mathbf{z})), \tag{3.59}$$

holds, where $\mathbf{k}(\mathbf{z})$ *is any occupation number that fulfils* $\underline{\zeta}(\mathbf{k}(\mathbf{z})) = \mathbf{z}$, *and*

2. for any occupation number $\mathbf{k}$ *we have*

$$P(\mathbf{K}=\mathbf{k}) = \binom{d}{\zeta_0(\mathbf{k}) \ldots \zeta_n(\mathbf{k})}^{-1} P(\mathbf{Z}=\underline{\zeta}(\mathbf{k})). \tag{3.60}$$

Remark 3.2.6.
(a) Under the assumption that the cells and the particles are indistinguishable, the possible configuration random variables may be identified with a strict sub-simplex of the simplex $M^1_{+,\mathrm{sym}}(\{1,\ldots,d\}^n)$ of symmetric probability measures.
(b) Under the assumption that the occupation number random variables are interchangeable, the factorial moments of the occupancy numbers can be determined by means of Lemma 3.2.6 and their marginals by Lemma 3.2.7. We do not reformulate these results here but rather consider some obvious consequences.

Lemma 3.2.8. *Assume that the occupation number random variables* $\mathbf{K}$ *are interchangeable, then*

1. for any $i, 0 \le i \le n$,

$$\begin{aligned} Var(Z_i/d) &= P(K_1=i, K_2=i) - P^2(K_1=i) \\ &+ \frac{1}{d}\{P(K_1=i) - P(K_1=i, K_2=i)\}, \end{aligned} \tag{3.61}$$

and

2. for $0 \le i, j \le n, i \ne j$,

$$\begin{aligned} Cov\left(\frac{Z_i}{d}, \frac{Z_j}{d}\right) &= P(K_1=i, K_2=j) - P(K_1=i)\,P(K_1=j) \\ &-\frac{1}{d}P(K_1=i, K_2=j). \end{aligned} \tag{3.62}$$

Remark 3.2.7. Obviously one can, besides the familiar three levels that are provided by the configurations, the occupation numbers and the occupancy numbers, define further levels by iteration of the scheme. For example, given an occupancy number $\mathbf{z}$, the forth level is defined by a vector $\underline{\rho} \in \{0,\ldots,n+1\}^{d+1}$ where

$$\rho_i(\mathbf{z}) = \sum_{m=0}^{n} \delta_{z_m,i}, \quad \sum_{i=0}^{d} \rho_i(\mathbf{z}) = n+1, \quad \sum_{i=0}^{d} i\,\rho_i(\mathbf{z}) = d. \tag{3.63}$$

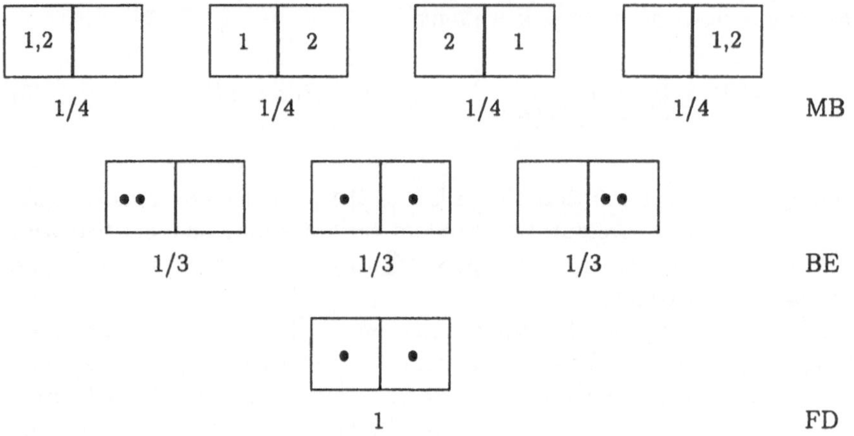

Fig. 3.1. Particles in boxes: traditional viewpoint

3.2.2 Statistics

Traditionally the three fundamental probability distributions (MB/BE/FD statistics) and their properties as well as their differences are introduced in a scheme that – for the special case $n = 2, d = 2$ – is shown in figure 3.1.

Here the three lines of cells characterize the events for the three statistics and the numbers below the cells are the probabilities. The numbers in the cells of the first line are the particle names. According to the traditional assumption these do exist for MB statistics ('distinguishable' or 'classical' particles) but not for BE statistics and FD statistics ('indistinguishable' or 'quantum' particles).

The fundamental disadvantage of this traditional scheme is that here two levels of the description are mixed. For MB statistics configurations are used whereas for BE and FD statistics occupation numbers are used. Moreover, it is confusing that for the three statistics there seem to be three different sets of events.

The scheme we propose instead consists of three levels which are shown – for the special case considered here – in figure 3.2.

In this scheme the left cell is cell 1 and the right cell is cell 2. In the upper line there are the configurations and the numbers in the cells are the particle names. Below the cells are the probabilities. Indistinguishability of the particles means that the probabilities of event $[1, 2]$ and event $[2, 1]$ agree. In the middle line there are the occupation numbers and the numbers in the cells are the numbers of particles in the cell. Indistinguishability of the cells means that the probabilities of event $[2, 0]$ and event $[0, 2]$ agree. The numbers in the boxes (these are not the cells) in the lower line are the occupancy numbers.

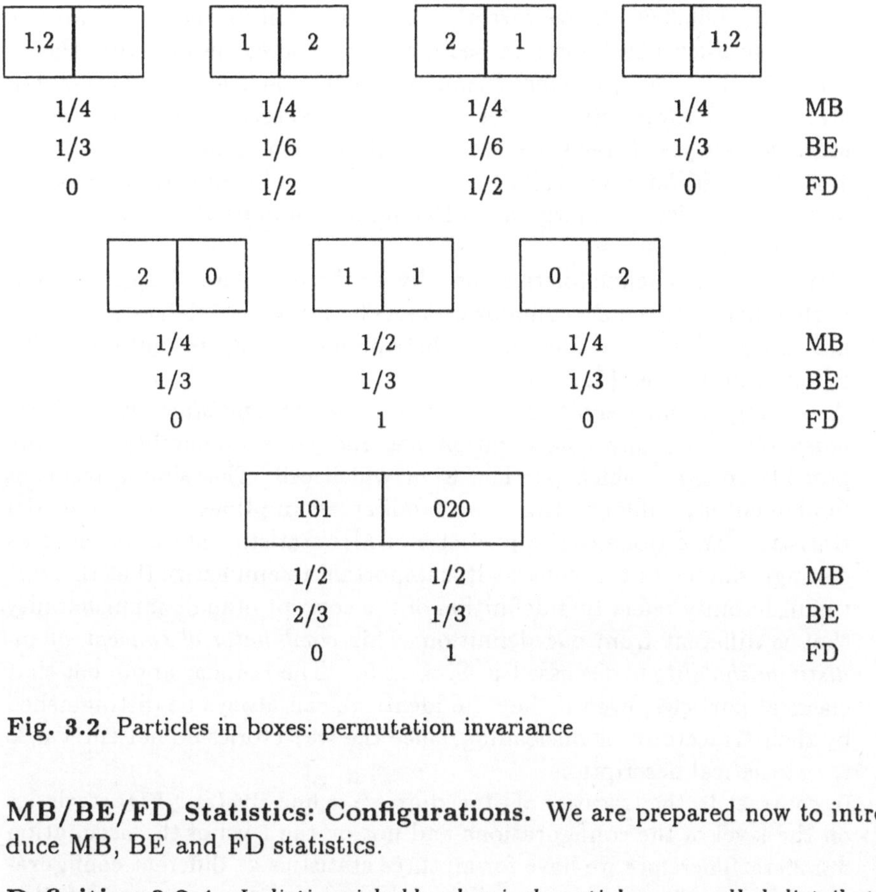

Fig. 3.2. Particles in boxes: permutation invariance

MB/BE/FD Statistics: Configurations. We are prepared now to introduce MB, BE and FD statistics.

Definition 3.2.4. *Indistinguishable classical particles are called distributed according to*

1. Maxwell-Boltzmann statistics *if*

$$P_{MB}(\mathbf{J}=\mathbf{j}) = \left(\frac{1}{d}\right)^n, \tag{3.64}$$

2. Bose-Einstein statistics *if*

$$P_{BE}(\mathbf{J}=\mathbf{j}) = \begin{pmatrix} n \\ \kappa_1(\mathbf{j}) \ \dots \ \kappa_d(\mathbf{j}) \end{pmatrix}^{-1} \begin{pmatrix} d+n-1 \\ n \end{pmatrix}^{-1}, \tag{3.65}$$

3. Fermi-Dirac statistics *if* $d \geq n$ *and*

$$P_{FD}(\mathbf{J}=\mathbf{j}) = \begin{cases} \frac{1}{n!}\begin{pmatrix} d \\ n \end{pmatrix}^{-1} & \text{if } \mathbf{k} \in \{0,1\}^d, \\ 0 & \text{otherwise.} \end{cases} \tag{3.66}$$

Remark 3.2.8.

(a) Eqs.(3.64,3.65,3.66) follow directly from the quantum theory of indistinguishable d-level systems (see Section 2.3). Whereas in quantum theory (for d-level systems) bosons, fermions and the particles of quantum MB statistics are three *sets* of states, in the classical setting MB, BE and FD statistics are just three states (probability distributions).

(b) Eqs.(3.64,3.65,3.66) are valid for fixed n and d. The same formulae hold, however, also for the marginals if the number of particles is reduced (see Section 3.3).

(c) MB statistics is defined here as usually by the postulate of equal *a priori* probabilities on the d^n configurations. The random variables $J_1, \ldots, J_n$ are independent and identically distributed and J_1 is uniformly distributed on the set $\{1, \ldots, d\}$.

(d) According to our definition *the particles of MB statistics are indistinguishable.* Since any configuration has the same probability it is impossible to infer which particle is in which cell. This simple result is fundamentally different from the familiar assumptions concerning MB statistics. Traditionally the particles of MB statistics are considered as distinguishable. In this context it is important to emphasize that this distinguishability refers to a definition of the concept of indistinguishability that is different from our definition. This *combinatorial concept of indistinguishability* is discussed in Section 5.2. The familiar argument that classical particles, even if they are identical, can always be distinguished by their trajectories is misleading, since the trajectories do not enter into the statistical description.

(e) In contrast to the traditional procedure we define BE (and FD) statistics on the level of the configurations and not on the level of the occupation numbers. Therefore we have for all three statistics d^n different configurations, but configurations that differ from one another by a permutation of the particles have the same probability and cannot be distinguished by their probabilities. Obviously, in our formalism for FD statistics configurations and occupation numbers with more than one particle in a cell exist, but these events have probability zero. In our setting the exclusion principle is a probabilistic assumption (an assumption concerning the probabilities of certain events) and not an ontological assumption concerning the existence of certain events. Therefore the unextendibility of the particles of FD statistics is a genuine probabilistic property.

(f) BE statistics can be defined by the postulate of equal *a priori* probabilities on the $\binom{d+n-1}{n}$ different occupation numbers.

(g) FD statistics is characterized by two properties. First, the *exclusion principle* holds, that is, $P_{FD}(\mathbf{J} = \mathbf{j}) = 0$ if there is a cell which contains more than one particle. Second, excluding these events with probability zero, FD statistics is the uniform distribution on the remaining $d_{[n]}$ configurations.

(h) Our strategy to define the familiar three statistics on the level of the configurations has the following advantage. First, these definitions follow immediately from the quantum theory of indistinguishable d-level systems. Second, we obtain a unified formalism for the three statistics. Third, we can determine the correlations of the particles of BE and FD statistics. This is impossible in the traditional formalism where, based on the assumption of an 'ontological identity' of events that differ from one another by a permutation of the particles, the description is confined to occupation numbers so that joint probabilities for the configurations (particles) do not exist.

(j) Obviously MB statistics and BE statistics are mixtures of normal statistics and parastatistics whereas FD statistics is a mixture of parastatistics.

MB/BE/FD Statistics: Occupation Numbers and Occupancy Numbers. Equivalently – using Theorem 3.2.1 – these probability distributions can be introduced on the level of the occupation numbers by means of the probability distributions

$$P_{MB}(\mathbf{K}=\mathbf{k}) = \binom{n}{k_1 \dots k_d} (\frac{1}{d})^n, \tag{3.67}$$

$$P_{BE}(\mathbf{K}=\mathbf{k}) = \binom{d+n-1}{n}^{-1}, \tag{3.68}$$

$$P_{FD}(\mathbf{K}=\mathbf{k}) = \begin{cases} \binom{d}{n}^{-1} & \text{if } \mathbf{k} \in \{0,1\}^d \\ 0 & \text{otherwise.} \end{cases} \tag{3.69}$$

Remark 3.2.9.

(a) The occupation number random variables of the three statistics are interchangeable such that, in this case, the cells and the particles are indistinguishable.

(b) Eqs.(3.67,3.68,3.69) are valid for fixed n and d. The same formulae hold, however, also for the marginals if the number of particles is reduced (see Section 3.3).

Since the occupation number random variables are interchangeable for the familiar three statistics we obtain, using Theorem 3.2.2, for the occupancy numbers

$$P_{MB}(\mathbf{Z}=\mathbf{z}) = \binom{d}{z_0 \dots z_n} \frac{n!}{0!^{z_0} \dots n!^{z_n}} (\frac{1}{d})^n, \tag{3.70}$$

$$P_{BE}(\mathbf{Z}=\mathbf{z}) = \binom{d}{z_0 \dots z_n} \binom{d+n-1}{n}^{-1}, \tag{3.71}$$

$$P_{FD}(\mathbf{Z}=\mathbf{z}) = \begin{cases} 1 & \text{if } \mathbf{z} = (d-n, n, 0, \dots, 0) \\ 0 & \text{otherwise.} \end{cases} \tag{3.72}$$

Remark 3.2.10.
(a) Eqs.(3.70,3.71) are due to BOLTZMANN 1877 (see Section 5.1). The most probable values for the occupation numbers and occupancy numbers in MB statistics and BE statistics are determined by VON MISES [149, 150].
(b) Whereas for MB and FD statistics the occupancy number random variables are not interchangeable, for BE statistics these random variables are interchangeable and max$(n+1)$–extendible.

This concludes the introduction of the familiar three statistics. Since the probability distributions (3.64,3.65, 3.66) describe classical indistinguishable particles the traditional distinction between MB statistics ('classical' statistics) and BE and FD statistics ('quantum' statistics) describing distinguishable and indistinguishable particles, respectively, becomes meaningless and the emphasis is shifted to more relevant physical properties of these particles, namely to their correlations (which are determined in Section 3.3).

3.2.3 Groups

In applications it is sometimes necessary to divide the cells into groups which are specified by some additional property of the particles, for example by their (discrete) energies, and to introduce, instead of the detailed (fine grained) information given by the occupation numbers $\mathbf{K}$, the (coarse grained) information concerning the number of particles in each group.

Definition. For the description we assume that a partition $\mathbf{D}$ of $\{1,\ldots,d\}$ is given. A partition $\mathbf{D} = \{D_1,\ldots,D_f\}, 2 \le f \le d$, of $\{1,\ldots,d\}$ is a family of non-void mutually disjoint subsets $D_i, 1 \le i \le f$, of $\{1,\ldots,d\}$ such that $\bigcup D_i = \{1,\ldots,d\}$ holds. The number of cells in group D_i is denoted by $d_i = |D_i|, 1 \le i \le f$, such that $\sum d_i = d$ holds. If instead of the complete information $\mathbf{D}$ only the number of cells in each group $\mathbf{d} = (d_1,\ldots,d_f)$ is relevant we say that the partition is specified by the vector $\mathbf{d}$.

For any partition $\mathbf{D}$ and any occupation number $\mathbf{k}$ we define the *group occupation number* $\mathbf{n}(\mathbf{k}) \in \{0,1,\ldots,n\}^f$ by, $1 \le i \le f$,

$$n_i(\mathbf{k}) = \sum_{m \in D_i} k_m, \tag{3.73}$$

such that

$$\sum_{i=1}^{f} n_i(\mathbf{k}) = n \tag{3.74}$$

holds.

Definition 3.2.5. *For any sequence of occupation number random variables* $\mathbf{K}$ *and any partition* $\mathbf{D}$ *of the cells the random variables* $N_i : \Omega \to \{0,1,\ldots,d\}, 1 \le i \le f$, *defined by*

$$N_i = \sum_{m \in D_i} K_i \tag{3.75}$$

are called group occupation number random variables.

Remark 3.2.11.

(a) According to the definition, the group occupation number random variables are subject to the deterministic constraint

$$\sum_{i=1}^{f} N_i = n, \quad \text{P-a.e.} \tag{3.76}$$

(b) The definition implies that

$$P(\mathbf{N} = \mathbf{n}) = \sum_{k_i, i \in D_1} \cdots \sum_{k_i, i \in D_f} P(\mathbf{K} = \mathbf{k}), \tag{3.77}$$

where each summation extends over those $k_i, i \in D_j$, which satisfy $\sum k_i = n_j, 1 \leq j \leq f$.

Two Levels of Indistinguishability.

Lemma 3.2.9. *Assume that the occupation number random variables* **K** *are interchangeable, then*

1. *for* $1 \leq i \leq f$,

$$EN_i = d_i \frac{n}{d}, \tag{3.78}$$

and

2. *for* $1 \leq i, j \leq f, i \neq j$,

$$Cor(N_i, N_j) = -\sqrt{\frac{\frac{d_i}{d}\frac{d_j}{d}}{(1-\frac{d_i}{d})(1-\frac{d_j}{d})}}. \tag{3.79}$$

Proof. Property (i) follows from the definition. Property (ii) follows from

$$E(N_i N_j) = d_i\, d_j\, E(K_1 K_2), \tag{3.80}$$

$$\mathrm{Var}(N_i) = d_i \,\mathrm{Var}(K_1) + d_i\,(d_i - 1)\,\mathrm{Cov}(K_1, K_2) \tag{3.81}$$

and eq.(3.50). QED

This lemma shows that the correlation coefficient (3.79), well–known from the multinomial distribution, is the general result of two levels of interchangeability, and holds whenever the cells and the particles are indistinguishable. For more details and the marginal distributions of the group occupation random variables **N** we refer to Section 3.3.

MB/BE/FD Statistics. From Theorem 3.2.2 we obtain for the familiar three statistics the following distributions

$$P_{MB}(\mathbf{N}=\mathbf{n}) = \binom{n}{n_1 \dots n_f} \prod_{i=1}^{f} \left(\frac{d_i}{d}\right)^{n_i}, \tag{3.82}$$

$$P_{BE}(\mathbf{N}=\mathbf{n}) = \binom{d+n-1}{n}^{-1} \prod_{i=1}^{f} \binom{d_i+n_i-1}{n_i}, \tag{3.83}$$

$$P_{FD}(\mathbf{N}=\mathbf{n}) = \begin{cases} \binom{d}{n}^{-1} \prod_{i=1}^{f} \binom{d_i}{n_i} & \text{if } n_i \leq d_i, 1 \leq i \leq f \\ 0 & \text{otherwise.} \end{cases} \tag{3.84}$$

We conclude this section with the remark that group occupancy numbers and *group occupancy number random variables* for BE statistics have been introduced by BOSE [35] in 1924. The most probable values of these random variables (subject to two constraints) are analyzed, for example, in [149, 109]. Explicit expressions for the distribution of the group occupancy numbers, however, are not established in the literature.

3.3 Brillouin Statistics

In this section we consider a special class of indistinguishable classical particles. This family of probability distributions, generalizing MB, BE and FD statistics, was introduced by BRILLOUIN [37, 38] in 1927. From the mathematical viewpoint Brillouin statistics is a special family of multivariate Pólya distributions (see for example [132, 97]). We therefore call these distributions Pólya–Brillouin distributions. In the first part of this section these distributions are introduced by means of a non–markovian stochastic process, the Pólya process. Next, by means of a related Markov process, group occupation numbers and occupancy numbers are considered. Finally, limit laws for Brillouin statistics are analyzed.

3.3.1 The Pólya Process

We assume that n identical particles, d cells and a partition $\mathbf{d}$ of the cells are given and derive the probability distribution of the group occupation numbers by means of a stochastic process. In any step of this process one particle is distributed onto one of the groups according to a conditional probability that depends upon the history of the process. This picturesque description is not necessarily thought to describe a (discretized) time-dependent physical process, but is essentially considered as a convenient model to derive Brillouin statistics and to understand its properties.

Our notation is as follows. For any partition, specified by $\mathbf{d} = (d_1, \ldots, d_f)$, $2 \leq f \leq d$, we introduce the *group configuration random variables* $G_i : \Omega \to \{1, \ldots, f\}, 1 \leq i \leq n$, defined on (Ω, F, P), where $[G_i = j], 1 \leq i \leq n, 1 \leq j \leq f$, denotes the event that particle i is in some cell of the jth group of cells. Whenever we are interested into the cells themselves (and not into the groups of cells), that is $\mathbf{d} = (1, \ldots, 1)$, we use instead of $\mathbf{G}$ the notation $\mathbf{J}$ of the last section.

A stochastic process $(G_1, \ldots, G_n)$ is uniquely determined by the initial condition $P(G_1 = j), 1 \leq j \leq f$, and the conditional probabilities $P(G_i = j | G_1 = j_1, \ldots G_{i-1} = j_{i-1}), 1 \leq j \leq f, (j_1, \ldots, j_{i-1}) \in \{1, \ldots, f\}^{i-1}, 2 \leq i \leq n$. By iteration this yields the family of joint distributions, $1 \leq m \leq n$,

$$\begin{aligned} &P(G_i = j_i, 1 \leq i \leq m) \qquad (3.85) \\ &= P(G_1 = j_1) \prod_{i=2}^{m} P(G_i = j_i | G_1 = j_1, \ldots, G_{i-1} = j_{i-1}) \end{aligned}$$

Moreover, for $1 \leq i \leq n$ and $1 \leq j \leq f$ we define the number $n_j(i)$ of particles in the cells of the jth group after i steps of the process by

$$n_j(i) = \sum_{m=1}^{i} \delta_{j, j_m}, \qquad (3.86)$$

so that

$$\sum_{j=1}^{f} n_j(i) = i \tag{3.87}$$

is the total number of particles which have been distributed after i steps. We remark that $n_j(i)$ depends explicitly upon the first i components of the configuration $\mathbf{j}$. For $n_j(n)$ we write n_j and for the special partition $\mathbf{d} = (1, \ldots, 1)$ we write $k_j(i)$ instead of $n_j(i)$.

Definition 3.3.1. *The stochastic process* $(G_1, \ldots, G_n), G_i : \Omega \to \{1, \ldots, f\}$, *determined by*

1. *the partition* $\mathbf{d} = (d_1, \ldots, d_f)$ *of the cells,*
2. *a real number* c*, the parameter of the process,*
3. *the initial condition*
$$P(G_1 = j) = d_j/d, \tag{3.88}$$
and
4. *iteratively by the conditional probabilities,* $1 \leq i \leq n-1$,
$$\begin{aligned} &P(G_{i+1} = j \,|\, G_1 = j_1, \ldots, G_i = j_i) \\ &\quad = \frac{d_j + c\, n_j(i)}{d + c\, i} P(G_1 = j_1, \ldots, G_i = j_i), \end{aligned} \tag{3.89}$$

is called Pólya process.

Remark 3.3.1.
(a) This definition makes sense for $c \geq 0$. For strictly negative values of the parameter c it is not *a priori* obvious whether the iteration of the conditional probabilities is compatible both with the non-negativity of probabilities and with the specified number of particles. In the following we confine ourselves throughout, as far as strictly negative values of c are concerned, to the set

$$c \in C_- = \{x \in \mathbf{R}_- ; x = -1/r, r \in \mathbf{N}\}. \tag{3.90}$$

Non-negativity of any probability requires $rd_j \geq n_j, 1 \leq j \leq f$. This entails $rd \geq i, 1 \leq i \leq n$, so that the process necessarily stops at most after $n = rd$ steps. Moreover, for configurations non-negativity requires $r \geq k_j, 1 \leq j \leq d$, so that each cell cannot accommodate more than r particles. The defining eq.(3.89) makes sense for $c \in C_-$ even if the conditional probabilities

$$P(G_{i+1} = j | G_1 = j_1, \ldots, G_i = j_i) = \frac{d_j + c\, n_j(i)}{d + c\, i} \tag{3.91}$$

are negative, as long as $rd > i$ is satisfied (the condition is fulfilled if we assume that $n \leq rd$ holds). This distinguished behaviour is due to the multiplicative structure of the joint probabilities (3.85) where any probability zero – that is reached if $n_j(i) = rd_j$ is attained – is translated into the future of the process without violating both the non-negativity of the probabilities and their normalization. It is the structure of the set of parameters C_- which guarantees this behavior. For other negative values of c the process necessarily stops before reaching probability zero. According to the results of [156] for $c < 0, c = -1/r, r > 0$, the sequence is $\max(N^*)$-extendible, where $N^* = [\min\{|d_i r|; 1 \leq i \leq f\}]$.

(b) Since the conditional probabilities (3.91) depend, via $n_j(i)$, explicitly on the history of the process, the Pólya process is non-markovian unless $c = 0$.

Lemma 3.3.1. *For $c \geq 0$ or $c \in C_-$ (under the assumption $rd \geq n, rd_j \geq n_j, 1 \leq j \leq f$,) the joint probability of the group configuration random variables $\mathbf{G} = (G_1, \ldots, G_n)$ is given by*

$$P(\mathbf{G} = \mathbf{j}) = \frac{\prod_{i=1}^{f} d_{j_i}(d_{j_i} + c)(d_{j_i} + 2c) \cdots (d_{j_i} + (n_{j_i} - 1)c)}{d(d+c)(d+2c) \cdots (d + (n-1)c)}$$
$$= \{(d/c)^{[n]}\}^{-1} \prod_{i=1}^{f} (d_{j_i}/c)^{[n_{j_i}]}, \tag{3.92}$$

where for $x \in \mathbf{R}$ and $n \in \mathbf{Z}_+$ the ascending factorials *are defined by*

$$x^{[n]} = x(x+1) \cdots (x+n-1) = \Gamma(x+n)/\Gamma(x). \tag{3.93}$$

Proof. This follows by an iteration of eq.(3.89) where for $c \in C_-$ we assume that $rd \geq n, rd_j \geq n_j, 1 \leq j \leq f$. QED

Remark 3.3.2.

(a) Eq.(3.92) depends only upon the number of particles n_i in the ith group, $1 \leq i \leq f$, after n steps and shows that the random variables $\mathbf{G}$ are interchangeable. This is due to the irrelevance of the order in which the particles are distributed onto the groups of cells, and in our context this implies that the particles of Brillouin statistics are indistinguishable.

(b) By construction and interchangeability the m-dimensional marginal distribution of $(G_1, \ldots, G_n)$, $1 \leq m < n$, is given by

$$P(G_1 = j_1, \ldots, G_m = j_m) = \{(d/c)^{[\tilde{n}]}\}^{-1} \prod_{i=1}^{f} (d_{j_i}/c)^{[\tilde{n}_{j_i}]}, \tag{3.94}$$

where for $1 \leq i \leq f$

$$\tilde{n}_i = \sum_{\ell=1}^{m} \delta_{i,j_\ell}, \tag{3.95}$$

and

$$\tilde{n} = \sum_{i=1}^{f} \tilde{n}_i. \tag{3.96}$$

(c) The familiar three statistics are easily recovered. For $c = 0$ Brillouin statistics corresponds to MB statistics whereas for $c = 1\,(c = -1)$ it is equivalent to BE (FD) statistics. For the special choices where $c \in C_-$ these statistics are known as *intermediate statistics* (see for example [76]); they constitute a natural generalization of FD statistics since for $c = -1/r, r \in \mathbf{N}$, any cell accommodates at most r particles.

(d) Let us, for the moment, forget about the interpretation of the parameter $\mathbf{d}$ as a partition and assume that $\mathbf{d} = (d_1, \ldots, d_f) \in \mathbf{R}_+^f$ such that $\sum_i d_i = d \in \mathbf{R}_+^f$. Then for $c \neq 0$ the scaling property

$$P(G_{i+1} = j \mid G_1 = j_1, \ldots, G_i = j_i) = \frac{d_j + cn_j(i)}{d + ci} = \frac{\frac{d_j}{c} + n_j(i)}{\frac{d}{c} + i} \tag{3.97}$$

holds. In particular, for $c = -1/r \in C_-$ we can replace any Brillouin statistics with partition $\mathbf{d} = (d_1, \ldots, d_f)$ by FD statistics with the enlarged partition $\mathbf{d}' = (rd_1, \ldots, rd_f)$ and for $c > 0$ we can replace any Brillouin statistics with partition $\mathbf{d} = (d_1, \ldots, d_f)$ by BE statistics with the refined partition $\mathbf{d}' = (d_1/c, \ldots, d_f/c)$.

Lemma 3.3.2. *The correlation coefficient $Cor(G_1, G_2)$ of Brillouin statistics (whenever $P(G_1 = j_1, G_2 = j_2)$ is well defined) is given by*

$$Cor(G_1, G_2) = \frac{c}{d + c} = \frac{1}{\frac{d}{c} + 1}. \tag{3.98}$$

Proof. Because of $P(G_1 = j) = d_j/d$ we have

$$EG_1 = \sum_{j=1}^{f} j\frac{d_j}{d}, \quad EG_1^2 = \sum_{j=1}^{f} j^2\frac{d_j}{d}. \tag{3.99}$$

Next, with

$$P(G_1 = j, G_2 = j') = \begin{cases} \frac{d_j d_{j'}}{d(d+c)} & \text{if } \; j \neq j', \\ \frac{d_j + c}{d+c}\frac{d_j}{d} & \text{else,} \end{cases} \tag{3.100}$$

we obtain

$$EG_1G_2 = \frac{d}{d+c}(\,EG_1\,)^2 + \frac{c}{d+c}EG_1^2. \tag{3.101}$$

QED

Remark 3.3.3.

(a) This correlation coefficient is independent of the number of groups. In particular we have

$$\mathrm{Cor}_{BE}(G_1, G_2) = \frac{1}{d+1} \quad , \quad \mathrm{Cor}_{FD}(G_1, G_2) = -\frac{1}{d-1}. \tag{3.102}$$

Under the assumption that d is large enough, for $c > 0$ the particles tend to bunch in those cells where particles are already present (positive correlations – bunching), whereas for $c < 0$ the particles try to avoid those cells where particles are already present (negative correlations – antibunching).

(b) For $c \geq 0$ the interchangeable random variables $\mathbf{G}$ are ∞-extendible by construction. Combining the correlation inequality (3.14) with the correlation coefficient (3.98) we once again obtain that for $c \in C_-$ the process is at most $\max(rd)$–extendible. By construction is is obvious that the group configuration random variables $\mathbf{G}$ are $\max(rd)$–extendible.

(c) For configurations

$$P(J_{i+1} = j | J_1 = j_1, \ldots, J_i = j + i) \propto 1 + c\, k_j(i) \tag{3.103}$$

is the conditional probability that cell j accommodates particle $i+1$. The right hand side of eq.(3.103) is well-known from EINSTEIN's theory of the emission and absorption of light quanta [57]. The probability that a cell is occupied is equal to to the probability that a photon is emitted. For statistically independent photons (MB statistics) the unnormalized conditional probability is proportional to one and the process corresponds to the spontaneous emission of a photon. In thermal equilibrium (BE statistics), however, there is an additional term which is proportional to the intensity $k_j(i)$, that is, to the number of photons already emitted (or the cells already occupied). This term corresponds to the stimulated emission of a photon.

3.3.2 The Number Process

The occupation number random variables of groups $\mathbf{N} = (N_1, \ldots, N_f)$ are determined by the indicator random variables $1 \leq i \leq f$,

$$N_i = \sum_{m=1}^{n} 1_{[G_m = i]}, \tag{3.104}$$

and subject to the deterministic constraint

$$\sum_{i=1}^{f} N_i = n \quad \text{P-a.e.} \tag{3.105}$$

Since there are $\begin{pmatrix} n \\ n_1 \ldots n_f \end{pmatrix}$ possibilities to distribute n particles onto f groups such that the jth group contains n_j particles we obtain, using Theorem

3.2.1, the probability distribution of the group occupation numbers from eq.(3.92).

Definition 3.3.2. *For any partition specified by* **d** *and any* $c \geq 0$ *or* $c \in C_-$ *the probability distributions of the group occupation number random variables*

$$P(\mathbf{N}=\mathbf{n}) = \binom{n}{n_1 \ldots n_f} \{(d/c)^{[n]}\}^{-1} \prod_{i=1}^{f} (d_i/c)^{[n_i]} \tag{3.106}$$

are called Pólya–Brillouin distributions.

Remark 3.3.4.
(a) Eq.(3.106) also holds for $c = 0$ (see eq.(3.108) below).
(b) For $c = -1/r \in C_-$ eq.(3.106) is defined only if $n \leq rd$. In this case $P(\mathbf{N}=\mathbf{n}) = 0$ if $n_j > rd_j$ for some $j, 1 \leq j \leq f$.
(c) The group occupation numbers **N** in Brillouin statistics are in general not interchangeable, whereas the occupation number random variables **K**, specified by the partition $\mathbf{d} = (1, \ldots, 1)$,

$$P(\mathbf{K}=\mathbf{k}) = \binom{n}{k_1 \ldots k_d} \{(d/c)^{[n]}\}^{-1} \prod_{i=1}^{d} (1/c)^{[k_i]}, \tag{3.107}$$

are interchangeable so that in Brillouin statistics both the particles and the cells are indistinguishable.
(d) We summarize some special cases of Pólya–Brillouin distributions.
 1. For $c = 0$ we obtain the multinomial distribution (3.82)

$$P_{MB}(\mathbf{N}=\mathbf{n}) = M_{n,(d_1/d,\ldots,d_f/d)}(\mathbf{n}). \tag{3.108}$$

 2. For $c > 0$ eq.(3.106) can be rewritten as (compare, for BE statistics, eq.(3.83))

$$P(\mathbf{N}=\mathbf{n}) = \binom{\frac{d}{c}+n-1}{n}^{-1} \prod_{i=1}^{f} \binom{\frac{d_i}{c}+n_i-1}{n_i}. \tag{3.109}$$

 3. For $c = -1/r \in C_-$ we obtain a polyhypergeometric distribution (compare, for FD statistics, eq.(3.84))

$$P(\mathbf{N}=\mathbf{n}) = \binom{rd}{n}^{-1} \prod_{i=1}^{f} \binom{rd_i}{n_i}. \tag{3.110}$$

(e) The expectation and the correlation coefficient are given by eqs.(3.78,3.79). For Brillouin statistics the variance is given by, $1 \leq i \leq f$,

$$\mathrm{Var}(N_i) = E(N_i)\ \{1 + \frac{c}{d_i}\ E(N_i)\}\ (\frac{1-\frac{d_i}{d}}{1-\frac{c}{d}}). \tag{3.111}$$

For $c \neq 0$ the Pólya process for the group configuration random variables is no Markov process. Considering, however, the group occupation numbers after each step of the Pólya process as a stochastic process, $\mathbf{N}(1), \ldots \mathbf{N}(n)$, it turns out that this process is a Markov process (see e.g. [62]).

Lemma 3.3.3. *The Markov process* $\mathbf{N}(i) : \Omega \to \{\mathbf{j} \in \{0, 1, \ldots, i\}^f; \sum_{m=1}^{i} j_m = i\}, 1 \leq i \leq n$, *defined by*

1. *the initial condition,* $1 \leq j \leq f$,

$$P(\mathbf{N}(1) = \mathbf{e}(j)) = d_j/d, \tag{3.112}$$

where $\mathbf{e}(j) \in \{0,1\}^f, e_i(j) = \delta_{i,j}$, *and*

2. *the transition probabilities,* $1 \leq i \leq n-1$,

$$P(\mathbf{N}(i+1) = \mathbf{n}(i) + \mathbf{e}(j) \,|\, \mathbf{N}(i) = \mathbf{n}(i)) = \frac{d_j + c\, n_j(i)}{d + c\, i}, \tag{3.113}$$

determines the Pólya-Brillouin distributions.

Proof. We obtain

$$P(\mathbf{N}(n) = \mathbf{n}(n)) = \sum_{\mathbf{n}(1),\ldots,\mathbf{n}(n-1)} P(\mathbf{N}(n) = \mathbf{n}(n), \mathbf{N}(n-1) = \mathbf{n}(n-1), \ldots, \mathbf{N}(1) = \mathbf{n}(1)). \tag{3.114}$$

Since $\mathbf{n}(n) = \sum_{j=1}^{f} n_j(n)\, \mathbf{e}(j)$, the sum is equal to the sum over all paths from some $\mathbf{e}(j), 1 \leq j \leq f$, to $\mathbf{n}(n)$, so that this process corresponds to a random walk on a $f$dimensional lattice. There are altogether $\begin{pmatrix} n \\ n_1 \ldots n_f \end{pmatrix}$ paths and, by virtue of interchangeability, any path has the same probability, given by eq.(3.92). QED

Remark 3.3.5.

(a) This random walk has been analyzed in [114] and studied, for BE statistics, detail in [30].

(b) Since for any $j, 1 \leq j \leq f$, and for any $i, 1 \leq i \leq n-1$,

$$E\left(\frac{d_j + cN_j(i+1)}{d + c(i+1)} \,\Big|\, \frac{d_j + cN_j(i)}{d + ci}\right) = \frac{d_j + cN_j(i)}{d + ci} \tag{3.115}$$

holds, any component of the proportions $(\mathbf{d} + c\mathbf{N}(i))/(d + ci)$ is a martingale (see for example [62, 30]).

From the viewpoint of physics one of the most important properties of the Pólya–Brillouin distributions is their compatibility (invariance) under any coarse graining of the partition $\mathbf{d}$.

Theorem 3.3.1. *Assume that the indices* $(1, \ldots, f)$ *of* $\mathbf{d}$ *are partitioned into* $g, 1 < g < f$, *groups* $(\tilde{D}_1, \ldots, \tilde{D}_g)$ *which are non-empty, mutually exclusive and exhaustive and define the occupation numbers of the coarse grained groups by*

$$\tilde{N}_i = \sum_{j \in \tilde{D}_i} N_j, \tag{3.116}$$

then, for $\mathbf{n} \in \{0, \ldots, n\}^g, \sum n_i = n$,

$$P(\tilde{\mathbf{N}} = \mathbf{n}) = \binom{n}{n_1 \quad \cdots \quad n_g} \{(\frac{d}{c})^{[n]}\}^{-1} \prod_{i=1}^{g} (\frac{\tilde{d}_i}{c})^{[n_i]} \tag{3.117}$$

holds, where

$$\tilde{d}_i = \sum_{j \in \tilde{D}_i} d_j. \tag{3.118}$$

Proof. We have,

$$P(\tilde{\mathbf{N}} = \mathbf{n}) = \sum_{n'_j, j \in \tilde{D}_1} \cdots \sum_{n'_j, j \in \tilde{D}_g} P(\mathbf{N} = \mathbf{n}'), \tag{3.119}$$

where the ith sum extends over those n'_j such that $\sum n'_j = n_i$ is satisfied. Inserting eq.(3.106) we obtain

$$\begin{aligned} P(\tilde{\mathbf{N}} = \mathbf{n}) & \qquad (3.120) \\ = \sum_{n'_j, j \in \tilde{D}_1} \cdots \sum_{n'_j, j \in \tilde{D}_g} \binom{n}{n'_1 \quad \cdots \quad n'_f} \{(\frac{d}{c})^{[n]}\}^{-1} \prod_{i=1}^{f} (\frac{d_i}{c})^{[n'_i]} \\ = \{(\frac{d}{c})^{[n]}\}^{-1} n! \prod_{i=1}^{g} \sum_{n'_j, j \in \tilde{D}_i} \prod_{j \in \tilde{D}_i} \frac{1}{n'_j!} (\frac{d_j}{c})^{[n'_j]}. \end{aligned}$$

Since for any $i, 1 \leq i \leq g$

$$\sum_{n'_j, j \in \tilde{D}_i} n_i! \prod_{j \in \tilde{D}_i} (\frac{d_j}{c})^{[n_j]} \frac{1}{n'_j!} = (\frac{\tilde{d}_i}{c})^{[n_i]} \tag{3.121}$$

holds, where

$$\tilde{d}_i = \sum_{j \in \tilde{D}_i} d_j, \quad n_i = \sum_{j \in \tilde{D}_i} n'_j, \tag{3.122}$$

the assertion follows. QED

Remark 3.3.6.

(a) Marginals of the group occupation numbers $\mathbf{N}$ are distributed according to a Pólya–Brillouin distribution. In particular, for the partition $\mathbf{d} = (1, \ldots, 1)$, the marginals of the occupation numbers $\mathbf{K}$ are determined by a Pólya–Brillouin distribution.

(b) The probability $P(N = k), 0 \leq k \leq n$, of the occupation number of the special partition $\mathbf{d} = (1, d-1), f = 2$, is identical to the probability $P(K = k)$ of the event that an arbitrarily chosen cell contains exactly k particles, $0 \leq k \leq n$,

$$P(K = k) = \binom{n}{k} \{(\frac{d}{c})^{[n]}\}^{-1} (\frac{1}{c})^{[k]} (\frac{d-1}{c})^{[n-k]}, \quad (3.123)$$

$$P_{MB}(K = k) = B_{n, \frac{1}{d}}(k), \quad (3.124)$$

$$P_{BE}(K = k) = \binom{d+n-1}{n}^{-1} \binom{d+n-k-2}{n-k}, \quad (3.125)$$

$$P_{FD}(K = k) = B_{1, \frac{n}{d}}(k). \quad (3.126)$$

For sake of completeness we remark that the distribution of the occupancy numbers in Brillouin statistics may be evaluated from the general theory by means of eq.(3.59). The formula for MB statistics ($c = 1$) is eq.(3.70). For $c > 0$ we obtain (compare eq.(3.71))

$$P(\mathbf{Z} = \mathbf{z}) = \{\frac{1}{d!} \binom{\frac{d}{c} + n - 1}{n}\}^{-1} \prod_{i=1}^{n} \frac{1}{z_i!} \binom{\frac{1}{c} + i - 1}{n}^{z_i}, \quad (3.127)$$

and for $c \in C_-, c = -1/r$ we have (compare eq.(3.72))

$$P(\mathbf{Z} = \mathbf{z}) = \begin{cases} 0 & \text{if} \quad z_i \neq 0 \quad \text{for} \quad i > r, \\ \{\frac{1}{d!} \binom{rd}{n}\}^{-1} \prod_{i=1}^{n} \frac{1}{z_i!} \binom{r}{i}^{z_i} & \text{otherwise.} \end{cases} \quad (3.128)$$

3.3.3 Macroscopic Limit and Continuum Limit I

Macroscopic Limit. In applications of the statistical scheme one usually is interested in macroscopic and continuous quantities. Therefore we analyze in this section limit laws of Brillouin statistics. The analysis of the LLN is deferred to Section 4.1 and Section 4.2. The *macroscopic limit* of Brillouin statistics is defined as follows.

Theorem 3.3.2. *In the limit $n, d \to \infty$, such that $n/d \to \overline{n} \in \mathbf{R}_+$ where $d_i, 1 \leq i \leq f-1$, remain fixed and the f-th group acts as a reservoir ($n_f \to \infty, d_f \to \infty$) the multivariate Pólya–Brillouin distributions $P(\mathbf{N} = \mathbf{n})$ factorize for $c \geq 0$ in general, and for $c = -1/r \in C_-$ under the condition $(rd - n) \to \infty$*

$$\lim_{n,d\to\infty} P(\mathbf{N}=\mathbf{n}) = \prod_{i=1}^{f-1} P(\mathcal{M}_i = n_i), \tag{3.129}$$

where the distribution of the random variables $\mathcal{M}_i$ is defined as follows.

1. *For $c = 0$ we obtain a Poisson distribution, $k \in \mathbf{Z}_+$,*

$$P(\mathcal{M}_i = k) = \pi_{d_i \overline{n}}(k). \tag{3.130}$$

2. *For $c > 0$ we obtain a negative binomial distribution, $k \in \mathbf{Z}_+$,*

$$P(\mathcal{M}_i = k) = \binom{\frac{d_i}{c} + k - 1}{k} \left(\frac{1}{1+c\overline{n}}\right)^{\frac{d_i}{c}} \left(\frac{c\overline{n}}{1+c\overline{n}}\right)^k. \tag{3.131}$$

3. *For $c = -1/r, r \in \mathbf{N}$ we obtain a binomial distribution, $0 \le k \le r\, d_i$,*

$$P(\mathcal{M}_i = k) = B_{r\, d_i, \overline{n}/r}(k). \tag{3.132}$$

Proof. (1) is a consequence of the classical Poisson limit theorem (cf. e.g. [62] and [97, Chapter 6.3.1.3.]). For (2) cf. e.g. [62, Chapter V.8. example 24] and [97, Chapter 6.3.1.2]. For (3) we observe that the condition $(rd - n) \to \infty$ guarantees that $\overline{n}/r \in [0,1]$. Cf. e.g. [97, Chapter 6.3.1.1.]. QED

Remark 3.3.7.

(a) Obviously, one group of cells can arbitrarily be chosen for the reservoir. For notational convenience we have chosen the fth group.

(b) According to this theorem the group occupation numbers N_i converge in distribution to the *macroscopic group occupation numbers* $\mathcal{M}_i$. For the special partition $\mathbf{d} = (1, \ldots, 1, d_f)$ the *macroscopic occupation numbers* are denoted by $\mathcal{K}_i$. The most important aspect of the macroscopic limit is the asymptotic statistical independence of the group occupation numbers, occupation numbers, respectively. Moreover, the macroscopic occupation numbers $\mathcal{K}_i$ are identically distributed and therefore denoted by $\mathcal{K}$.

(c) Eq.(3.130) includes, for MB statistics, the Poisson distribution ($d_i = 1$)

$$P_{MB}(\mathcal{K} = k) = \pi_{\overline{n}}(k), \tag{3.133}$$

eq.(3.131), for BE statistics, the geometric distribution ($d_i = 1$)

$$P_{BE}(\mathcal{K} = k) = \frac{1}{1+\overline{n}} \left(\frac{\overline{n}}{1+\overline{n}}\right)^k, \tag{3.134}$$

and eq.(3.132), for FD statistics, the binomial distribution ($d_i = 1$)

$$P_{FD}(\mathcal{K} = k) = B_{1,\overline{n}}(k). \tag{3.135}$$

(d) From eqs.(3.130) – (3.132) we obtain the expectations and the variances, $1 \leq i \leq f-1$,

$$c = 0 \quad : \quad E(\mathcal{M}_i) = d_i\,\overline{n}, \quad \mathrm{Var}(\mathcal{M}_i) = d_i\,\overline{n}, \tag{3.136}$$

$$c > 0 \quad : \quad E(\mathcal{M}_i) = d_i\,\overline{n}, \quad \mathrm{Var}(\mathcal{M}_i) = d_i\,\overline{n}\,(1 + c\,\overline{n}), \tag{3.137}$$

$$c < 0 \quad : \quad E(\mathcal{M}_i) = d_i\,\overline{n}, \quad \mathrm{Var}(\mathcal{M}_i) = d_i\,\overline{n}\,(1 - \frac{\overline{n}}{r}). \tag{3.138}$$

Therefore, for all values of the parameter c the functional relation (compare the limit of eq.(3.111))

$$\mathrm{Var}(\mathcal{M}_i) = E(\mathcal{M}_i)\,(1 + \frac{c}{d_i}\,E(\mathcal{M}_i)). \tag{3.139}$$

holds.

(e) For BE statistics eq.(3.139) is EINSTEIN's famous fluctuation equation

$$\mathrm{Var}_{BE}(\mathcal{M}_i) = E_{BE}(\mathcal{M}_i) + \frac{1}{d_i} E^2_{BE}(\mathcal{M}_i), \tag{3.140}$$

consisting of the 'particle' term $E_{BE}(\mathcal{M}_i)$ and the 'wave' term $E^2_{BE}(\mathcal{M}_i)/d_i$. This interpretation is analyzed in Section 5.1. On the other hand, for FD statistics we have

$$\mathrm{Var}_{FD}(\mathcal{M}_i) = E_{FD}(\mathcal{M}_i) - \frac{1}{d_i} E^2_{FD}(\mathcal{M}_i). \tag{3.141}$$

(f) A random variable $\mathcal{M}$ with values in $\mathbf{Z}_+$ has *super-Poisson statistics* or *sub-Poisson statistics* if

$$\mathrm{Var}(\mathcal{M}) > E(\mathcal{M}) \quad \text{or} \quad \mathrm{Var}(\mathcal{M}) < E(\mathcal{M}) \tag{3.142}$$

holds. Accordingly, the macroscopic group occupation numbers $\mathcal{M}$ of Brillouin statistics have super-Poisson statistics for $c > 0$ and sub-Poisson statistics for $c < 0, c \in C_-$.

'Particle' Limit. Allowing, in a further step, the number of cells in each group to grow to infinity while the mean particle number $\overline{n}$ decreases to zero, provides us with the *'particle' limit.*

Lemma 3.3.4. *(For $c > 0$ see [62, Chapter XI.6. example (a)].) Under the assumptions of Theorem 3.3.2 suppose moreover that $\overline{n} \to 0, d_i \to \infty, 1 \leq i \leq f-1$, such that $d_i\,\overline{n} \to \overline{n}_i \in \mathbf{R}_+, 1 \leq i \leq f-1$. Then for all $c \geq 0$ and for all $c \in C_-$ the distribution of the macroscopic group occupation numbers converge to a Poisson distribution*

$$\lim_{d_i \to \infty, \overline{n} \to 0} P(\mathcal{M}_i = k) = \pi_{\overline{n}_i}(k). \tag{3.143}$$

Remark 3.3.8. This lemma follows also from a general Poisson limit theorem for interchangeable configuration random variables under the assumption $n, d, d_i \to \infty, d_i\,n/d \to \overline{n}_i$, such that d grows faster than n (see [40, Theorem 2.1.2.]).

'Wave' Limit. Letting grow, on the other hand, the mean particle number $\overline{n}$ to infinity and using the scaled macroscopic group occupation numbers $\varepsilon\mathcal{M}_i$ where $\varepsilon > 0$ (which is interpreted usually as an energy) decreases to zero, we obtain the *macroscopic continuum limit.*

Lemma 3.3.5. *Under the assumptions of Theorem 3.3.2 suppose moreover that $\overline{n} \to \infty, \varepsilon \to 0$, such that $\varepsilon E(\mathcal{M}_i) = \varepsilon d_i \overline{n} \to d_i \kappa \in \mathbf{R}_+$ (this implies $\varepsilon \overline{n} \to \kappa$). Then for $c \geq 0$ the scaled macroscopic group occupations $\varepsilon\mathcal{M}_i$ converge in distribution to non-negative random variables $\mathcal{E}_i$.*

1. *For $c = 0$ we obtain*
$$\mathcal{E}_i = d_i \kappa \qquad P\text{-a.e.} \tag{3.144}$$
2. *For $c > 0$ the random variable $\mathcal{E}_i$ is distributed according to a Γ-distribution, $x \geq 0$,*
$$\Gamma_{\alpha,\beta}(x) = \frac{1}{\Gamma(\beta)} \alpha^\beta x^{\beta-1} \exp(-\alpha x), \tag{3.145}$$
with scale parameter α and parameter β where
$$\alpha = \frac{1}{c\kappa} \quad \text{and} \quad \beta = \frac{d_i}{c}. \tag{3.146}$$
3. *For $c < 0, c \in C_-$ the limit does not exist.*

Proof. We use the equivalence of convergence in distribution with the convergence of the associated characteristic functions.

(1) For $c = 0$ we have
$$\begin{aligned} E\exp(it\varepsilon\mathcal{M}_i) &= \exp\{d_i\overline{n}(e^{it\varepsilon} - 1)\} \\ &= \exp\{d_i\overline{n}\varepsilon \frac{(e^{it\varepsilon} - 1)}{\varepsilon}\} \to \exp(itd_i\kappa). \end{aligned} \tag{3.147}$$

(2) For $c > 0$ we have
$$E\exp(it\varepsilon\mathcal{M}_i) = \{1 - c\overline{n}(e^{it\varepsilon} - 1)\}^{-d_i/c} \tag{3.148}$$
$$= \{1 - c\overline{n}\varepsilon \frac{e^{it\varepsilon} - 1}{\varepsilon}\}^{-d_i/c} \to (1 - c\kappa\, it)^{-d_i/c}.$$

(3) For $c = -1/r < 0$ we obtain
$$\begin{aligned} E\exp(it\varepsilon\mathcal{M}_i) &= \{1 + \frac{\overline{n}}{r}(e^{it\varepsilon} - 1)\}^{d_i r} \\ &= \{1 + \frac{\overline{n}\varepsilon}{r} \frac{e^{it\varepsilon} - 1}{\varepsilon}\}^{d_i r} \to (1 + \frac{it\kappa}{r})^{d_i r}. \end{aligned} \tag{3.149}$$

This, however, is not the characteristic function of a probability measure because of the following argument. From part (2) we know that
$$f_-(t) = (1 - c\kappa\, it)^{-d_i/c} \tag{3.150}$$

is the characteristic function of $\mathcal{E}_i$ for $c > 0$. Accordingly

$$f_+(t) = (1 + c\kappa\, it)^{-d_i/c} \tag{3.151}$$

is the characteristic function of $-\mathcal{E}_i$ for $c > 0$. Setting $c = 1/r$ reveals that

$$f_+(t) = (1 + \frac{\kappa}{r}\, it)^{-d_i r} \tag{3.152}$$

is a characteristic function. Therefore

$$\frac{1}{f_+(t)} = (1 + \frac{\kappa}{r}\, it)^{d_i r} \tag{3.153}$$

cannot be a characteristic function since (see [113]) both $f(t)$ and $1/f(t)$ are characteristic functions iff $f(t) = \exp(itx)$. QED

Remark 3.3.9.

(a) For BE statistics this limit is usually called the *'wave' limit* (see the discussion in Section 5.1). It is a well-known fact that the continuum limit of the geometric distribution is the exponential distribution (cf. e.g. [63])

$$\Gamma_{\frac{1}{\kappa},1}(x) = \frac{1}{\kappa}\exp(-x/\kappa). \tag{3.154}$$

(b) For strictly negative values of the parameter c, and in particular for FD statistics, no continuum limit exists because of the unextendibility of the underlying configuration random variables.

Macroscopic Occupancy Numbers. In the macroscopic limit the normalized occupancy number random variables become deterministic quantities.

Lemma 3.3.6. *In the macroscopic limit of Brillouin statistics for any $c \geq 0$ and any $c \in C_-$, the random variables Z_k/d converge for any $k \in \mathbf{Z}_+$ in distribution to a random variable $\mathcal{Z}_k$ which is equal to a constant*

$$\mathcal{Z}_k = P(\mathcal{K} = k) \quad P\text{-a.e.} \tag{3.155}$$

Proof. By virtue of eq.(3.61) we have

$$\begin{aligned}\lim_{n,d\to\infty} \mathrm{Var}(Z_k/d) = \lim_{n,d\to\infty} [\, & P(K_1 = k, K_2 = k) - P^2(K_1 = k) \\ + & \frac{1}{d}\{P(K_1 = k) - P(K_1 = k, K_2 = k)\}\,].\end{aligned} \tag{3.156}$$

Since in the macroscopic limit both $P(K_1 = k) = o(1)$ and $P(K_1 = k, K_2 = k) = o(1)$ hold, and due to the asymptotic independence of the occupation numbers

$$\lim_{n,d\to\infty} P(K_1 = k, K_2 = k) = \lim_{n,d\to\infty} P^2(K_1 = k), \tag{3.157}$$

the limit under consideration is P-a.e. a constant and therefore equal to the mean value. This one, however, is equal to $P(\mathcal{K} = k)$. QED

Remark 3.3.10.

(a) In the macroscopic limit the expectation of the normalized occupancy number Z_k/d is equal to the most probable value. Analogously, eq.(3.62) may be used to show that the normalized occupancy number random variables Z_k/d are asymptotically independent.

(b) For MB statistics these results are due to VON MISES [150]. Almost all results by VON MISES on MB statistics extend without modification to the occupancy numbers of Brillouin statistics with $c > 0$.

4. De Finetti's Theorem

In this chapter, which methodologically is the center of our investigation, we derive de Finetti's representation theorem for probabilities of ∞-extendible interchangeable random variables. We confine ourselves in general to a discrete setting, that is to a system of indistinguishable particles which are distributed onto d groups of cells. We first derive de Finetti's theorem following de Finetti's original proof. In a second step we generalize this theorem to a multivariate context. Finally, limit laws of de Finetti's theorem are established.

4.1 De Finetti's Classical Theorem

In this section we consider simplest system where particles are distributed on 2 groups of cells. For the description of the statistical scheme we use the interchangeable random variables $G_i : \Omega \to \{0,1\}, 1 \leq i \leq n$ where $[G_i = 1]$ denotes the event that particle i is in the first group of cells ($\mathbf{d} = (d_1, d - d_1)$) or in the first cell ($d = 2$).

4.1.1 De Finetti's Proof

The Finite de Finetti Formula.

Lemma 4.1.1. *[44]. Assume that the random variables $G_i : \Omega \to \{0,1\}, 1 \leq i \leq n$, are interchangeable. Then for any $m, 1 \leq m \leq n$, and for any $k, 0 \leq k \leq m$, de Finetti's formula for finite sequences of interchangeable random variables holds*

$$P(\sum_{i=1}^{m} G_i = k) = \sum_{r=k}^{n-m+k} \binom{n}{m}^{-1} \binom{r}{k} \binom{n-r}{m-k} P(\sum_{i=1}^{n} G_i = r). \tag{4.1}$$

Proof. Conditioning with respect to $\sum_{i=1}^{n} G_i$ and resummation yields

$$P(\sum_{i=1}^{m} G_i = k) = \sum_{r=0}^{n} P(\sum_{i=1}^{m} G_i = k \mid \sum_{i=1}^{n} G_i = r)\, P(\sum_{i=1}^{n} G_i = r). \tag{4.2}$$

The conditional probability is zero if $r < k$ or if $r - k > n - m$ since the condition cannot be satisfied in these cases. On the other hand for $k \le r \le n - m + k$,

$$P(\sum_{i=1}^{m} G_i = k \mid \sum_{i=1}^{n} G_i = r) = \frac{P(\sum_{i=1}^{n} G_i = r, \sum_{i=1}^{m} G_i = k)}{P(\sum_{i=1}^{n} G_i = r)}. \tag{4.3}$$

There are $\binom{m}{k}$ possibilities to select k particles from the set of first m particles for the first group. For each of these possibilities there are $\binom{n-m}{r-k}$ possibilities to select $r - k$ particles for the first group from the remaining $n - m$ particles. By virtue of interchangeability the probabilities of all these choices agree and are equal, for example, to $P(G_1 = 1, \dots, G_r = 1, G_{r+1} = 0, \dots, G_n = 0)$. Therefore,

$$\frac{P(\sum_{i=1}^{n} G_i = r, \sum_{i=1}^{m} G_i = k)}{P(\sum_{i=1}^{n} G_i = r)} = \tag{4.4}$$

$$\frac{\binom{m}{k}\binom{n-m}{r-k} P(G_1 = 1, \dots, G_r = 1, G_{r+1} = 0, \dots, G_n = 0)}{\binom{n}{r} P(G_1 = 1, \dots, G_r = 1, G_{r+1} = 0, \dots, G_n = 0)}$$

$$= \binom{n}{r}^{-1}\binom{m}{k}\binom{n-m}{r-k} = \binom{n}{m}^{-1}\binom{r}{k}\binom{n-r}{m-k}.$$

QED

Remark 4.1.1.
(a) The distribution of $\sum_{i=1}^{m} G_i$ is a compound hypergeometric distribution, that is, a discrete mixture of hypergeometric distributions, $k \le \min(r, m)$,

$$\mathcal{H}_{n,m,r}(k) = \binom{n}{r}^{-1}\binom{m}{k}\binom{n-m}{r-k}. \tag{4.5}$$

Notice that $\mathcal{H}_{n,m,r} = \mathcal{H}_{n,r,m}$.
(b) For a more condensed formulation of eq.(4.1) we introduce for all m, $1 \le m \le n$, the occupation number of group one

$$N(m) = \sum_{i=1}^{m} G_i \tag{4.6}$$

and the *random measure* (this is a random variable) $\mathcal{H}_{n,m,N(n)}(\cdot)$. With this notation eq.(4.1) can be rewritten as (see also [99])

$$P(N(m) = k) = E(\mathcal{H}_{n,m,N(n)}(k)). \tag{4.7}$$

More explicitly we have

$$P(N(m)=k) = \int_0^1 dP^{N(n)}(r)\mathcal{H}_{n,m,r}(k). \tag{4.8}$$

(c) The compounding distribution $P^{N(n)}$ is uniquely determined by the interchangeable random variables $G_i, 1 \leq i \leq n$.
(d) For $m = n$ in eq.(4.1) we obtain the representation of the elements of the simplex $M^1_{+,\mathrm{sym}}(\{1,\dots,n\}^2)$ in terms of its extreme points (special hypergeometric distributions, cf. eq.(3.45)).

The Limit. The finite de Finetti formula (4.1) holds for finite and ∞-extendible sequences of interchangeable random variables. We now assume that the sequence under consideration is ∞-extendible and analyze the limit of this formula for $n \to \infty$. A further reformulation of eq.(4.8) explains the strategy that is applied in the sequel. We substitute $p = r/n$ and obtain

$$P(\sum_{i=1}^{m} G_i = k) = \int_{\alpha_n}^{\beta_n} d\nu_n(p)\, \mathcal{H}_{n,m,np}(k), \tag{4.9}$$

where for any $n \in \mathbf{N}$ the measure ν_n is the image of P under the intensive collective random variable $Q(n)$ which is defined by, $n \in \mathbf{N}$,

$$Q(n) = \frac{N(n)}{n}, \tag{4.10}$$

and where the integration extends from $\alpha_n = k/n$ to $\beta_n = (n-m+k)/n$. It is our goal to take the limit $n \to \infty$ in eq.(4.9). To this end we need, besides the trivial limits of α_n and β_n, information on the convergence of both the sequence of probability measures ν_n and the integrand $\mathcal{H}_{n,m,np}(k)$ (here m and k are fixed and $p \in \{0, 1/n, 2/n, \dots, 1\}$). Following DE FINETTI's strategy [47], the convergence of the sequence ν_n is derived from the convergence of the sequence of random variables $Q(n)$.

Lemma 4.1.2. *The sequence of probability measures $\nu_n \in M^1_+([0,1])$ defined by $\nu_n = P^{Q(n)}, n \in \mathbf{N}$, converges weakly to a probability measure $\nu = P^Q \in M^1_+([0,1])$. Equivalently, the LLN holds for the random variables $G_i, i \in \mathbf{N}$,*

$$\frac{1}{n}\sum_{i=1}^{n} G_i \to Q, \tag{4.11}$$

where the convergence is in distribution.

Proof. (Cf. [44, 87]) By virtue of interchangeability

$$E\{Q(n) - Q(n+r))\}^2 = \frac{r}{n(n+r)}\{E(G_1^2) - E(G_1G_2)\} \tag{4.12}$$

holds, so that the sequence $Q(n)$ is a Cauchy sequence in mean-square. The Riesz-Fischer theorem asserts that the Hilbert space of equivalence classes $L^2(\Omega, F, P)$ is complete. Accordingly, there exists a random variable Q with values in $[0,1]$ which is P-a.e. uniquely determined so that $Q = \text{l.i.m.}\, Q(n)$ where l.i.m. denotes the limit in mean square. Since convergence in mean square implies convergence in distribution and that one is equivalent to the weak convergence of the associated probability measures, we obtain the result.

QED

It remains to analyze the convergence of the hypergeometric distribution. The binomial approximation of the hypergeometric distribution includes uniform convergence.

Lemma 4.1.3. *(Cf. e.g. [62, 49].) For all $m \in \mathbf{N}$ and all $k, 0 \le k \le \min(m,r)$, in the limit $n, r \to \infty, r/n \to p \in [0,1]$, we have*

$$\lim_{n,r\to\infty} \sup_{p\,\in\,[0,1]} |\mathcal{H}_{n,m,r}(k) - B_{m,p}(k)| = 0. \tag{4.13}$$

Combining Lemma 4.1.2 and Lemma 4.1.3, we are now able to take the limit in eq.(4.9).

Theorem 4.1.1. de Finetti. *Assume that $G_i : \Omega \to \{0,1\}, i \in \mathbf{N}$, is a sequence of interchangeable random variables. Then there exists a uniquely determined probability measure $\nu \in M_+^1([0,1])$ such that for any $m \in \mathbf{N}$ and any $k, 0 \le k \le m$, the integral representation*

$$P(\sum_{i=1}^{m} G_i = k) = \int_0^1 d\nu(p) B_{m,p}(k) \tag{4.14}$$

holds. Moreover, the representing measure ν is determined by the LLN, i.e. $\nu = P^Q$ where

$$Q = \lim_{n\to\infty} \frac{1}{n} \sum_{i=1}^{n} G_i. \tag{4.15}$$

Proof. (Cf. [44, 87]) For any $m \in \mathbf{N}$ and any $k, 0 \le k \le m$, we have the estimate

$$\begin{aligned} &|\int_{\alpha_n}^{\beta_n} d\nu_n(p)\mathcal{H}_{n,m,pn}(k) - \int_0^1 d\nu(p) B_{m,p}(k)| \\ &\le |\int_{\alpha_n}^{\beta_n} d\nu_n(p)\mathcal{H}_{n,m,pn}(k) - \int_{\alpha_n}^{\beta_n} d\nu_n(p) B_{m,p}(k)| \\ &+ |\int_{\alpha_n}^{\beta_n} d\nu_n(p) B_{m,p}(k) - \int_0^1 d\nu(p) B_{m,p}(k)| \\ &= I_n^{(1)}(k) + II_n^{(2)}(k). \end{aligned} \tag{4.16}$$

For the first term we obtain

$$I_n^{(1)}(k) \leq \sup_{p \in [\alpha_n, \beta_n]} |\mathcal{H}_{n,m,np}(k) - B_{n,p}(k)| \tag{4.17}$$

which vanishes in the limit under consideration by virtue of Lemma 4.1.3. The assertion follows from the fact that $[\alpha_n, \beta_n] \to [0,1]$ and the weak convergence of the sequence ν_n (Lemma 4.1.2) since the binomial distribution, considered as a function of the parameter p, is continuous and bounded.

Next, we evaluate the characteristic function

$$\begin{aligned}\lim_{n\to\infty} E \exp(i\frac{t}{n}\sum_{i=0}^{n} G_i) &= \int d\nu(p) \sum_{k=0}^{n} B_{n,p}(k)\, e^{itk/n} \\ = \int d\nu(p)(1-p+pe^{it/n})^n &= \int d\nu(p)(1-p+p+\frac{itp}{n}+o(\frac{1}{n}))^n \\ = \int d\nu(p)(1+\frac{itp}{n})^n &= \int d\nu(p)e^{itp} = \hat{\nu}(t) = \widehat{P^Q},\end{aligned} \tag{4.18}$$

where, for the last step, we used Lemma 4.1.2. Uniqueness of the representing measure follows from the uniqueness of characteristic functions. QED

Remark 4.1.2.

(a) Following the proposal by [91] we call the mixing measure of a de Finetti-type theorem the *de Finetti measure*.

(b) In analogy to eq.(4.7) we reformulate the integral representation (4.14) by means of the random measure $B_{n,Q}(\cdot)$

$$P(N(n)=k) = E\, B_{n,Q}(k), \tag{4.19}$$

or more explicitly,

$$P(\sum_{i=1}^{n} G_i = k) = \int dP^Q(p) B_{n,p}(k). \tag{4.20}$$

Examples.

Example 4.1.1. MB Statistics. Assume that the particles are independent and identically distributed so that $P(G_1 = 1) = p \in [0,1]$ holds. We expect that the integral representation reproduces the binomial distribution. From eq.(4.15) we obtain

$$\begin{aligned}\lim_{n\to\infty} E\{\exp(it\frac{1}{n}\sum_{i=1}^{n} G_i)\} &= \lim_{n\to\infty}\sum_{k=0}^{n}\exp(it\frac{k}{n})\, B_{n,p}(k) \\ = \lim_{n\to\infty}\{1-p+p\exp(\frac{it}{n})\}^n &= \lim_{n\to\infty}\{1+p\frac{it}{n}+o(\frac{1}{n})\}^n.\end{aligned} \tag{4.21}$$

For $n \to \infty$ this characteristic function converges to $\exp(itp)$ which is the characteristic function of the Dirac measure concentrated at p

$$P_p(N(n) = k) = B_{n,p}(k) = \int d\delta_p(q)\, B_{n,q}(k). \tag{4.22}$$

Accordingly, the de Finetti measure is a Dirac measure iff the interchangeable random variables under consideration are independent and identically distributed.

Example 4.1.2. BE Statistics. For two cells and n particles the occupation number random variable K of BE statistics is uniformly distributed on the set $\{0, 1, \ldots, n\}$ so that

$$\begin{aligned}\lim_{n\to\infty} E\{\exp(it\,\frac{1}{n}\sum_{i=1}^{n} J_i)\} &= \lim_{n\to\infty}\sum_{k=0}^{n}\exp(it\frac{k}{n})\frac{1}{n+1}\\ = \lim_{n\to\infty}\frac{1}{1+n}\,\frac{1-\exp(it)\exp(\frac{it}{n})}{1-\exp(\frac{it}{n})} &= \frac{1-\exp(it)}{-it}.\end{aligned} \tag{4.23}$$

Accordingly, the de Finetti measure for the discrete uniform distribution on $\{0, 1, \ldots, n\}$ is the uniform distribution on $[0, 1]$

$$P_{BE}(K(n) = k) = \frac{1}{n+1} = \int_0^1 dp\, B_{n,p}(k). \tag{4.24}$$

Example 4.1.3. Brillouin Statistics. For $c \geq 0$ the interchangeable random variables of the Pólya process are ∞-extendible such that an integral representation for the occupation number for the first group of cells with d_1 cells exists. For $c > 0$ we have

$$\begin{aligned}&P(N(n) = k) =\\ &\binom{\frac{d}{c}+n-1}{n}^{-1}\binom{\frac{d_1}{c}+k-1}{k}\binom{\frac{d-d_1}{c}+n-k-1}{n-k}\\ &= \int_0^1 dp\; b_{d_1/c,(d-d_1)/c}(p)\, B_{n,p}(k),\end{aligned} \tag{4.25}$$

where for $\alpha > 0, \beta > 0$,

$$b_{\alpha,\beta}(p) = \frac{\Gamma(\alpha+\beta)}{\Gamma(\alpha)\,\Gamma(\beta)}\, p^{\beta-1}(1-p)^{\alpha-1} \tag{4.26}$$

is the beta-distribution. In particular, we obtain the integral representation for the marginal probability $P(K(n) = k)$ of the occupation number of one group in BE statistics, $f = 2, \mathbf{d} = (1, d-1)$,

$$\begin{aligned}&P_{BE}(K(n) = k) =\\ &\binom{d+n-1}{n}^{-1}\binom{d+n-k-2}{n-k} = \int dp\,(d-1)\,(1-p)^{d-2}\, B_{n,p}(k).\end{aligned} \tag{4.27}$$

The derivation of the de Finetti measure is omitted here since its is sufficient to show by direct computation that eq.(4.25) holds.

4.1.2 Remarks

A Proof by Choquet's Theory. The proof presented here follows the historical lines of argument [44]. A – now familiar – different proof, based upon martingale techniques (reversed martingales) is presented for example in [26].

A more abstract viewpoint is adopted by HEWITT and SAVAGE [90] who consider instead of a countably infinite sequence of interchangeable random variables with values in $\{0,1\}$ the convex set $\mathcal{M}_{\mathbf{N},2}$ of all symmetric probability measures on $\{0,1\}^{\mathbf{N}}$. It is an easy exercise to verify that the set of extreme points is given by the homogeneous product measures

$$\mathrm{ex}\,(\mathcal{M}_{\mathbf{N},2}) = \{\overset{\mathbf{N}}{\otimes}\mathcal{P};\mathcal{P} \in M^1_+(\{0,1\})\}. \tag{4.28}$$

According to CHOQUET's theory of barycentric representations of elements of convex compact sets for any $\Pi \in \mathcal{M}_{\mathbf{N},2}$ there exists a probability measure $\tilde{\nu}$ defined on the set of extreme points $\mathrm{ex}\,(\mathcal{M}_{\mathbf{N},2})$ such that the integral representation holds

$$\Pi = \int d\tilde{\nu}(\mathcal{P}) \overset{\mathbf{N}}{\otimes} \mathcal{P}. \tag{4.29}$$

Since $\mathcal{P}(\{1\}) = p$ for some $p \in [0,1]$, $\tilde{\nu}$ can be identified with a probability measure $\nu \in M^1_+([0,1])$. For the $n$dimensional marginal probabilities we eventually obtain, $\underline{\varepsilon} \in \{0,1\}^n$,

$$\Pi_n(\underline{\varepsilon}) = \int d\nu(p) \prod_{i=1}^{n} p^{\varepsilon_i}(1-p)^{1-\varepsilon_i} = \int d\nu(p)\, p^k\,(1-p)^{n-k}, \tag{4.30}$$

where $k = \sum \varepsilon_i$. This formula is once again de Finetti's representation theorem. For more information we refer the reader to [139, 93, 4].

The Strong Law of Large Numbers. In his fundamental memoir of 1930 [44] DE FINETTI obtained a result which is more general than Lemma 4.1.2, namely the strong LLN for the sequence $G_i, i \in \mathbf{N}$.

Theorem 4.1.2. *(Cf. [100, 45]). Assume that the random variables G_i : $\Omega \to \{0,1\}, i \in \mathbf{N}$, are interchangeable, then*

$$P(\lim_{n\to\infty} \frac{\sum_{i=1}^n G_i}{n} = Q) = 1. \tag{4.31}$$

Remark 4.1.3. In general, the classical limit theorems for i.i.d. random variables extend to ∞-extendible sequences of interchangeable random variables. For details we refer the reader to [4] and Section 4.3.

Correlation Inequalities. The integral representations provided by de Finetti's theorem can be used to establish correlation inequalities for ∞-extendible interchangeable random variables that do not hold for unextendible sequences. We give one example for an occupancy event. We consider the event $\sum_{j=1}^{d} P(J_1 = j, \ldots, J_n = j)$ corresponding to the special occupancy number $\mathbf{z} = (d-1, 0, \ldots, 0, 1)$ where all particles are confined to one cell.

Lemma 4.1.4. *Assume that the random variables $J_i : (\Omega, F, P) \to \{1, \ldots, d\}$, $1 \leq i \leq n$, are interchangeable and ∞-extendible, then for any $n \in \mathbf{N}$ the multiple maximal occupations are bounded from below*

$$\sum_{j=1}^{d} P(J_1 = j, \ldots, J_n = j) \geq (\frac{1}{d})^n, \tag{4.32}$$

where equality holds iff $P = P_{MB}$.

Proof. Hölder's inequality implies

$$\sum_{i=1}^{d} p_i^n \geq (\frac{1}{d})^n, \tag{4.33}$$

where equality holds iff $p_i = 1/d, 1 \leq i \leq d$. QED

Remark 4.1.4.
(a) For BE statistics the inequality is satisfied. For FD statistics the inequality is violated such that the configuration random variables for FD statistics are unextendible. For MB statistics equality holds such that MB statistics has the least possible weight on maximal multiparticle configurations.
(b) For Brillouin statistics with $c > 0$ the inequality is satisfied whereas for $c < 0, c \in C_-$ it is violated.

Conditional Independence. It is a characteristic feature of de Finetti's theorem that the integral representations are given in terms of probabilities of i.i.d. random variables. These belong to the basic concepts of probability theory and a modern version of de Finetti's theorem [109] states that a countably infinite sequence of random variables is interchangeable iff the random variables are conditionally independent and identically distributed.

Theorem 4.1.3. *(Cf. e.g. [109, 40]) Assume that the random variables $X_i, i \in \mathbf{N}, E|X_1| < \infty$, are interchangeable such that by virtue of the strong LLN $\underline{X} = \lim \sum X_i/n$ exists P-a.e., then the random variables X_i are i.i.d. conditionally upon $\underline{X}$, that is*

1. *for all $n \in \mathbf{N}$ and for all $x \in \mathbf{R}^n$*

$$P(X_i < x_i, 1 \leq i \leq n \,|\, \underline{X}) = \prod_{i=1}^{n} P(X_i < x_i \,|\, \underline{X}), \qquad (4.34)$$

2. *and for any choice of continuous functions $f_1, \ldots, f_n$, each one vanishing outside a finite interval,*

$$E\{f_1(X_1) \cdots f_n(X_n) \,|\, \underline{X}\} = \prod_{i=1}^{n} E(f_i(X_i) \,|\, \underline{X}) = \prod_{i=1}^{n} E(f_i(X_1) \,|\, \underline{X}). \qquad (4.35)$$

Conversely, if a sequence of random variables $X_i, i \in \mathbf{N}$, is i.i.d. conditionally upon the tail σ-field, then the sequence is interchangeable.

Remark 4.1.5.
(a) This theorem, due to LOÈVE [109], is the most important result concerning an infinite sequence of interchangeable random variables. It immediately implies de Finettis theorem.
(b) For technical details – in particular, the existence of a regular conditional distribution – we refer to [4, 40, 101].

4.1.3 The Physical Significance

From the Microdomain to the Macrodomain: LLN. From the physical viewpoint the normalized (group) occupation random variables $Q(n) = N(n)/n$ are intensive collective observables. We call the limit of this sequence the (normalized) *particle density random variable* Q. According to Lemma 4.1.2 for the sequence of interchangeable random variables $G_i, i \in \mathbf{N}$, the LLN holds which we reformulate here.

Lemma 4.1.5. *Assume that the random variables $G_i; \Omega \to \{0,1\}, i \in \mathbf{N}$, are interchangeable, then for any $f \in C_b([0,1])$*

$$\lim_{n\to\infty} E\, f(\frac{1}{n}\sum_{i=1}^{n} G_i) = E\, f(Q). \qquad (4.36)$$

Remark 4.1.6.
(a) It is a trivial fact that the microscopic properties of a system determine the macroscopic properties and one aspect of this general relationship between a microscopic system and its macroscopic version is expressed by the LLN (an other one, on a different scale, by the CLT). Notice, however, that in contrast to the familiar situation (i.i.d. random variables) the LLN for interchangeable variables entails no deterministic description for the macroscopic quantities: The (normalized) particle density Q is, by virtue of the particle correlations, in general a random variable.

(b) From Section 3.1 we recall that sequences of interchangeable random variables are not necessarily ∞-extendible. An example for unextendible sequences is Brillouin statistics with strictly negative values of the parameter $c, c \in C_-$, and, in particular, FD statistics. For these sequences the LLN cannot be formulated and, in turn, de Finetti's theorem is not applicable. Such systems are essentially finite systems and their properties are difficult to understand since an analysis of the convergence of the relative frequencies is impossible.

The integral representation of de Finetti's theorem has two aspects.

- The de Finetti measure is determined by the sequence of interchangeable random variables by virtue of the LLN.
- The de Finetti measure determines the distribution of the interchangeable random variables $G_i, i \in \mathbf{N}$.

From the Macrodomain to the Microdomain: de Finetti's Theorem. If $\nu = P^Q$ or Q is given, then for any $n \in \mathbf{N}$ and any $\underline{\varepsilon} \in \{0,1\}^n$

$$P(G_i = \varepsilon_i, 1 \le i \le n) = \int dP^Q(p)\, p^k (1-p)^{n-k}, \tag{4.37}$$

where $k = \sum \varepsilon_i$. In particular, all microscopic correlations are determined by the macroscopic observable Q. For the $n$particle correlation function we obtain

$$E \prod_{i=1}^{n} (G_i - E\, G_i) = E\,(Q - E\,Q)^n. \tag{4.38}$$

Given a description of a system on a macroscopic level (by means of the particle density) there are, in general, many different microscopic versions conceivable which yield, in the limit under consideration, the same macroscopic behavior. It is a general problem of any reductionistic strategy that reductionism is far from being unique. To fix the microscopic level we need, therefore, additional principles. Two such principles have been applied successfully and both are due to DE FINETTI

- ∞-extendible interchangeable random variables (permutation symmetry) and
- ∞-divisible laws (corresponding to limits of arrays of independent and identically distributed random variables).

Assuming that the particle density is known and that the particles are indistinguishable, we obtain by means of de Finetti's theorem a converse to the LLN: Let there be given a description of identical particles by means of the group configuration random variables $G_i, i \in \mathbf{N}$, of the statistical scheme and assume that the distribution of the particle density Q is known, then, imposing indistinguishability, the distribution of the group configuration random variables is uniquely determined. In other words: The microscopic (local) behaviour of a system of infinitely many indistinguishable particles is completely determined by the collective (global) properties of the system. This allows for a condensed formulation of all properties of the microscopic system in terms of collective observables.

4.2 The Multivariate de Finetti Theorem

In this section de Finetti's classical theorem is generalized to a multivariate situation. First Finetti's representation for probabilities of finite sequences of interchangeable random variables is extended and in the limit de Finetti's theorem is derived. Since this extension is straightforward we give, in general, no proofs. The multivariate de Finetti theorem is then used to establish a characterizing property of Brillouin statistics. Finally limit laws of Brillouin statistics are considered.

4.2.1 The Theorem

We use the notation of Section 4.1, that is, we consider the configuration random variables of groups $G_i : \Omega \to \{0, 1, \ldots, f\}, 1 \leq i \leq n$, for a partition $\mathbf{d}$ of the cells. For $1 \leq m \leq n$ we set $\mathbf{N}(m) = (N_1(m), \ldots, N_f(m))$ where, $1 \leq j \leq f$,

$$N_j(m) = \sum_{i=1}^{m} 1_{[G_i=j]} \tag{4.39}$$

denotes the group occupation number random variable of the jth group.

Lemma 4.2.1. *Assume that the random variables $G_i : \Omega \to \{0, 1, \ldots, f\}, 1 \leq i \leq n$, are interchangeable. Then for any $m, 1 \leq m \leq n$, and any $\mathbf{n} = \mathbf{n}(m) \in \{0, 1, \ldots, m\}^f, \sum n_i = m$, de Finetti's formula for finite sequences of interchangeable random variables holds*

$$P(\mathbf{N}(m) = \mathbf{n}) = \sum_{n_i \leq r_i \leq n-m-r_i} P(\mathbf{N}(n) = \mathbf{r}) \begin{pmatrix} n \\ m \end{pmatrix}^{-1} \prod_{i=1}^{f} \begin{pmatrix} r_i \\ n_i \end{pmatrix}, \tag{4.40}$$

where $\mathbf{r} = \mathbf{r}(n), \mathbf{r} \in \{0, 1, \ldots, n\}^f, \sum r_i = n$.

Remark 4.2.1.

(a) The distribution of the group occupation number random variables $\mathbf{N}(\mathrm{m})$ is a compound polyhypergeometric distribution. The extreme points of the simplex $M^1_{+,\mathrm{sym}}(\{1, \ldots, n\}^d)$ can be deduced from eq.(4.40) setting $m = n$.

(b) In the familiar picture (sampling without replacement), illustrating the polyhypergeometric distribution, the numbers r_i of balls with colour $i, 1 \leq i \leq f$, in the population of n balls are random variables.

For the derivation of an integral representation we follow the strategy of Section 4.1. Setting $p_i = r_i/n, 1 \leq i \leq f$, we obtain

$$P(\mathbf{N}(m) = \mathbf{n}) = \sum_{\mathbf{p}} P(\frac{\mathbf{N}(n)}{n} = \mathbf{p}) \begin{pmatrix} n \\ m \end{pmatrix}^{-1} \prod_{i=1}^{f} \begin{pmatrix} np_i \\ n_i \end{pmatrix}, \tag{4.41}$$

where $\alpha_i(n) \leq p_i \leq \beta_i(n), 1 \leq i \leq f$, and

$$\alpha_i(n) = \frac{n_i}{n}, \qquad \beta_i(n) = \frac{n - m + n_i}{n}. \tag{4.42}$$

In the multivariate setting the interval [0,1] is generalized to the set of probabilities

$$\mathcal{S}_f = \{\mathbf{p} \in \mathbf{R}^{f-1};\, p_i \geq 0, \sum p_i \leq 1\} \tag{4.43}$$

This set is a simplex which is contained in a $(f-1)$-dimensional hyperplane of $\mathbf{R}^f$. The volume of this simplex is given by the formula (cf. e.g. [67] or see eq.(2.131))

$$\text{Vol}(\mathcal{S}_f) = \int_{\mathcal{S}_f} dp_1 \cdots dp_{f-1} = \frac{1}{(f-1)!}. \tag{4.44}$$

In analogy to the one-dimensional case we define the vector-valued particle density random variable $\mathbf{Q}(n) : \Omega \to \mathcal{S}_f$ by

$$\mathbf{Q}(n) = \mathbf{N}(n)/n, \tag{4.45}$$

which is subject to the deterministic constraint

$$\sum_{i=1}^{f} Q_i(n) = 1, \qquad \text{P-a.e.} \tag{4.46}$$

Weak convergence of the associated induced probability measures is established in analogy to the one-dimensional case.

Lemma 4.2.2. *Assume that the random variables $G_i : \Omega \to \{1, \ldots, f\}, i \in \mathbf{N}$, are interchangeable, then the sequence of probability measures $\nu_n = P^{\mathbf{Q}(n)} \in M_+^1(\mathcal{S}_f)$ converges weakly to a probability measure $\nu = P^{\mathbf{Q}} \in M_+^1(\mathcal{S}_f)$ where $\mathbf{Q} : \Omega \to \mathcal{S}_f$ is the mean-square limit of the sequence of particle densities $\mathbf{Q}(n)$.*

Moreover, in the limit under consideration sampling without replacement converges to sampling with replacement.

Lemma 4.2.3. *For $m \in \mathbf{N}, \mathbf{n} \in \{0, 1, \ldots, m\}^f, \sum n_i = m$, in the limit $n \to \infty, r_i \to \infty, r_i/n \to p_i, 1 \leq i \leq f$,*

$$\lim_{n\to\infty} \sup_{\mathbf{p} \in \mathcal{S}_f} \left| \binom{n}{m}^{-1} \prod_{i=1}^{f} \binom{r_i}{n_i} - M_{m,\mathbf{p}}(\mathbf{n}) \right| = 0. \tag{4.47}$$

Combining the last two lemmata, we obtain the multivariate de Finetti theorem.

Theorem 4.2.1. *Assume that the random variables* $G_i : \Omega \to \{1,\ldots,f\}, i \in \mathbf{N}$, *are interchangeable, then there exists an uniquely determined probability measure* $\nu \in M_+^1(S_f)$ *such that for all* $n \in \mathbf{N}$ *and all* $\mathbf{n} \in \{0,1\ldots n\}^f, \sum n_i = n$, *the integral representation holds*

$$P(\mathbf{N}(n) = \mathbf{n}) = \int_{S_f} d\nu(\mathbf{p})\, M_{n,\mathbf{p}}(\mathbf{n}). \tag{4.48}$$

Moreover, the de Finetti measure ν *is determined by the formula,* $\mathbf{t} \in \mathbf{R}^{f-1}$,

$$\hat{\nu}(\mathbf{t}) = \int_{S_f} d\nu(\mathbf{p}) \exp(i\,\mathbf{t}\cdot\mathbf{p}) = \lim_{n\to\infty} E \exp\Big(\frac{i\,\mathbf{t}\cdot\mathbf{N}(n)}{n}\Big). \tag{4.49}$$

Remark 4.2.2.

(a) Using the particle density $\mathbf{Q}$ we can reformulate the integral representation (4.48)

$$P(\mathbf{N}(n) = \mathbf{n}) = \int_{S_f} dP^{\mathbf{Q}}(\mathbf{p})\, M_{n,\mathbf{p}}(\mathbf{n}). \tag{4.50}$$

(b) One aspect of de Finetti's theorem is the validity of the LLN for the interchangeable random variables $G_i, i \in \mathbf{N}$, that is,

$$\lim_{n\to\infty} E\, f\Big(\frac{\mathbf{N}(n)}{n}\Big) = f(\mathbf{Q}) \tag{4.51}$$

holds for any $f \in C_b(\mathbf{R}_+^{f-1})$.

Example 4.2.1. MB Statistics.

The de Finetti measure for the group occupation number random variables in MB statistics is the Dirac measure concentrated at the point $(d_1/d,\ldots,d_f/d) \in S_f$.

Example 4.2.2. BE Statistics.

Assume that the configuration random variables $J_i, i \in \mathbf{N}$, are distributed according to BE statistics, then

$$P_{BE}(\mathbf{K}(n) = \mathbf{k}) = \binom{d+n-1}{n}^{-1} = \int_{S_d} dp_1 \cdots dp_{d-1}\,(d-1)!\, M_{n,\mathbf{p}}(\mathbf{k}). \tag{4.52}$$

Remark 4.2.3.

(a) This formula is derived, for example, in [142, 104, 31].

(b) The de Finetti measure for BE statistics is the uniform distribution on the simplex S_d

$$d\nu(\mathbf{p}) = dp_1 \cdots dp_{d-1}\,(d-1)! \tag{4.53}$$

From the meaning of the LLN it is obvious that the discrete uniform distribution on the set of occupation numbers gives rise to a continuous uniform distribution.

(c) The de Finetti measure for BE statistics is derived in [30] by advanced methods (Martin boundary) from the special properties of the Markov process $\mathbf{K}(1), \ldots, \mathbf{K}(n)$. These techniques extend to the process $\mathbf{N}(1), \ldots, \mathbf{N}(n)$ for Brillouin statistics, $c \geq 0$.

4.2.2 The Dirichlet Distribution

Definition 4.2.1. *The probability distribution $\mathcal{D}_{\underline{\alpha}}$ on the simplex S_f where $\underline{\alpha} = (\alpha_1, \ldots, \alpha_f) \in \mathbf{R}_+^f, \alpha_i > 0, 2 \leq i \leq f, \sum \alpha_i = \alpha$,*

$$\mathcal{D}_{\underline{\alpha}}(\mathbf{p}) = \frac{\Gamma(\alpha)}{\prod_{i=1}^f \Gamma(\alpha_i)} \prod_{i=1}^f p_i^{(\alpha_i - 1)} \tag{4.54}$$

is called Dirichlet distribution

Remark 4.2.4.
(a) Since the components of $\mathbf{p} = (p_1, \ldots, p_f) \in S_f$ are not independent, here and in what follows the notation p_f always means $p_f = 1 - \sum_{i=1}^{f-1} p_i$.
(b) The definition makes sense if $\alpha_i \geq 0$ for all i and $\alpha_i > 0$ for some i. If $\alpha_i = 0$, the corresponding component is degenerate at zero.
(c) Assume that the random vector $\mathbf{Q} : \Omega \to S_f$ is distributed according to a Dirichlet distribution $D_{\underline{\alpha}}$. The moments then are given by the formula

$$E \prod_{i=1}^f Q_i^{r_i} = \frac{\prod_{i=1}^f \alpha_i^{[r_i]}}{\alpha^{[r]}}, \tag{4.55}$$

where $\mathbf{r} \in \mathbf{Z}_+^f, \sum r_i = r$. In particular, we have

$$E\,Q_i = \alpha_i/\alpha, \tag{4.56}$$

$$\mathrm{Var}(Q_i) = \frac{1}{1+\alpha} E\,Q_i\,(1 - E\,Q_i), \tag{4.57}$$

$$\mathrm{Cor}(Q_i, Q_j) = -\sqrt{\frac{\frac{\alpha_i}{\alpha}\frac{\alpha_j}{\alpha}}{(1 - \frac{\alpha_i}{\alpha})(1 - \frac{\alpha_j}{\alpha})}}. \tag{4.58}$$

(d) The marginals of a Dirichlet distribution are given by a Dirichlet distribution (see Theorem 4.2.3 below). Notice that the one-dimensional marginals are given by a beta-distribution (see eq.(4.25)).

Theorem 4.2.2. *(Cf. e.g. [7, 121].) Assume that the group configuration random variables of the statistical scheme are distributed according to Brillouin statistics, then for $c > 0$ these interchangeable random variables are ∞-extendible and according to de Finetti's theorem for the probability distribution of the group occupation number random variables an integral representation exists. The de Finetti measure is a Dirichlet distribution*

$$P(\mathbf{N} = \mathbf{n}) = \int_{S_f} dp_1 \cdots dp_{f-1} \mathcal{D}_{\mathbf{d}/c}(\mathbf{p})\, M_{n,\mathbf{p}}(\mathbf{n}). \tag{4.59}$$

Remark 4.2.5.

(a) The particle density random variables $\mathbf{Q}$ which are distributed according to the Dirichlet distribution $\mathcal{D}_{\mathbf{d}/c}$ have the following moments

$$\begin{aligned} E\,Q_i &= d_i/d, && (4.60) \\ \mathrm{Var}(Q_i) &= \frac{1}{1+\frac{d}{c}}\,E\,Q_i\,(1-E\,Q_i), && (4.61) \\ \mathrm{Cor}(Q_i,Q_j) &= -\sqrt{\frac{\frac{d_i}{d}\,\frac{d_j}{d}}{(1-\frac{d_i}{d})\,(1-\frac{d_j}{d})}}. && (4.62) \end{aligned}$$

This correlation coefficient is equivalent to the correlation coefficient (3.79) of the group occupation random variables, showing that for $c > 0$ the correlation structure is invariant under the LLN and, from the converse viewpoint, invariant under mixtures.

(b) In particular, for the occupation number random variables the correlation coefficient of the particle densities is given by, $d_i = 1, 1 \leq i \leq d$,

$$\mathrm{Cor}(Q_i,Q_j) = -\frac{1}{d-1}. \qquad (4.63)$$

An important property of the Dirichlet distribution is the compatibility under any coarse graining of the random variables described by this distribution.

Theorem 4.2.3. *(Cf. e.g. [64].) Assume that the random variables $Q_i : \Omega \to S_f, 1 \leq i \leq f$, are distributed according to a Dirichlet distribution $\mathcal{D}_{\underline{\alpha}}$, then for any partition of the indices $1,\ldots,f$ into $g, 1 < g < f$, nonempty mutually exclusive and exhaustive groups $I_1,\ldots,I_g$ the random variables $\tilde{Q}_j = \sum_{m\in I_j} Q_m, 1 \leq j \leq g$, are distributed according to a Dirichlet distribution $\mathcal{D}_{\underline{\tilde{\alpha}}}$ where*

$$\tilde{\alpha_j} = \sum_{m\in I_j} \alpha_m. \qquad (4.64)$$

Remark 4.2.6.

(a) Applied to the distribution of the particle densities $\mathbf{Q}$ this yields the compatibility under any coarse graining of the groups of cells (see the context of Theorem 3.3.1) where the particle densities of the coarse grained groups $\tilde{Q}_j = \sum_{m\in I_j} Q_m, 1 \leq j \leq g$, are distributed according to the Dirichlet distribution $\mathcal{D}_{\tilde{\mathbf{d}}/c}$ and $\tilde{\mathbf{d}}$ is given by the number of cells of the coarse grained groups

$$\tilde{d_j} = \sum_{m\in I_j} d_m. \qquad (4.65)$$

We conclude that the compatibility under coarse graining of the Pólya-Brillouin distributions, as shown in Theorem 3.3.1, remains valid on the level of the particle density random variables (LLN) or, from the viewpoint of the integral representation, is present already in the de Finetti measure.

(b) In Brillouin statistics both the particles and the cells are indistinguishable. On the level of the LLN this implies, for the special partition $\mathbf{d} = (1, \ldots, 1)$, that the particle density random variables $Q_i, 1 \leq i \leq d$ are interchangeable, such that whenever de Finetti's theorem applies, the de Finetti measure is a symmetric probability measure on $\mathcal{S}_d$. Accordingly, for $f = d, d_i = 1, 1 \leq i \leq d$, the Dirichlet distribution in eq.(4.59) is a symmetric distribution.

4.2.3 Continuum Limit and Macroscopic Limit II

Finally we consider the macroscopic limit of Brillouin statistics for $c \geq 0$ from the viewpoint of the de Finetti measure.

Theorem 4.2.4. *Assume that the particle density random variable* $\mathbf{Q} : \Omega \to \mathcal{S}_f$ *is distributed according to a Dirichlet distribution (4.54) with parameters* $\alpha_i = d_i/c, 1 \leq i \leq f$. *In the macroscopic continuum limit* $d \to \infty$ *such that* $d_1, \ldots d_{f-1}$ *remain fixed and the fth group acts as a reservoir, the random variable* $\kappa\, d\, \mathbf{Q}, \kappa > 0$, *converges in distribution to a random variable* $\underline{Q} : \Omega \to \mathbf{R}_+^{f-1}$. *The components of* $\underline{Q}$ *are statistically independent and* $Q_i : \Omega \to \mathbf{R}_+, 1 \leq i \leq f-1$, *is distributed according to a* Γ*-distribution with scale parameter* α *and parameter* β *where (cf. eqs.(3.145,3.146))*

$$\alpha = \frac{1}{c\kappa} \quad \textit{and} \quad \beta = \frac{d_i}{c}, \tag{4.66}$$

or, more explicitly,

$$\Gamma_{\frac{1}{c\kappa}, \frac{d_i}{c}}(x) = \frac{(\frac{1}{c\kappa})^{d_i/c}}{\Gamma(\frac{d_i}{c})} x^{(\frac{d_i}{c}-1)} \exp(-\frac{x}{c\kappa}). \tag{4.67}$$

Proof. In this proof the summations extend from $i = 1$ to $i = f-1$. We have, $\mathbf{t} \in \mathbf{R}^{f-1}$,

$$\begin{aligned} E \exp(i\,\kappa d\,\mathbf{t} \cdot \mathbf{Q}) &= \int_{\mathcal{S}_f} dp_1 \cdots dp_{f-1}\, \mathcal{D}_{\mathbf{d}/c}(\mathbf{p}) \exp(i\,\kappa d\,\mathbf{t} \cdot \mathbf{p}) \\ &= \int_{\kappa d \mathcal{S}_f} dx_1 \cdots dx_{f-1} \exp(i\,\mathbf{t} \cdot \mathbf{x}) \{\prod_{i=1}^{f-1} \frac{x_i^{(\frac{d_i}{c}-1)}}{\Gamma(\frac{d_i}{c})}\} \times \\ &\quad (\frac{1}{\kappa d})^{(\sum d_i/c)} \frac{\Gamma(\frac{d}{c})}{\Gamma(\frac{d-\sum d_i}{c})} \{1 - \frac{\sum x_i}{\kappa d}\}^{(\frac{d-\sum d_i}{c}-1)}. \end{aligned} \tag{4.68}$$

Since $\kappa d\, \mathcal{S}_f \to \mathbf{R}_+^{f-1}$ and

$$(\frac{1}{\kappa d})^{(\sum d_i/c)} \frac{\Gamma(\frac{d}{c})}{\Gamma(\frac{d-\sum d_i}{c})} = (\frac{1}{c\kappa})^{(\sum d_i/c)} \{1 + O(\frac{1}{d})\}, \tag{4.69}$$

we obtain

$$\lim_{d\to\infty} E \exp(i\kappa d\mathbf{t} \cdot \mathbf{Q}) = E \exp(i\mathbf{t} \cdot \underline{\mathcal{Q}}) \tag{4.70}$$

$$= \int_{\mathbf{R}^{f-1}} dx_1 \cdots dx_{f-1} \exp(i\mathbf{t} \cdot \mathbf{x}) \prod_{i=1}^{f-1} \frac{(\frac{1}{c\kappa})^{d_i/c}}{\Gamma(\frac{d_i}{c})} x_i^{(\frac{d_i}{c}-1)} \exp(-\frac{x_i}{c\kappa}).$$

QED

Remark 4.2.7.
(a) The collective random variables $\underline{\mathcal{Q}}$ are called *macroscopic particle densities.*
(b) For BE statistics and the partition $\mathbf{d} = (1, d_f - 1)$ we obtain the exponential distribution (cf. eq.(3.154))

$$\Gamma_{\frac{1}{\kappa},1}(x) = \frac{1}{\kappa} \exp(-x/\kappa). \tag{4.71}$$

Lemma 4.2.4. *(Cf. [121]) Assume that the random variables $\mathcal{Q}_i, 1 \leq i \leq f$, are independent and distributed according to the Γ-distributions (4.67), then the proportions $\mathbf{Q} : \Omega \to S_f$, defined by*

$$Q_i = \frac{\mathcal{Q}_i}{\sum_{i=1}^{f} \mathcal{Q}_i}, \tag{4.72}$$

are distributed according to the Dirichlet distribution $\mathcal{D}_{\mathbf{d}/c}$.

Remark 4.2.8.
(a) Here we recover another situation where microscopic properties are determined by collective macroscopic ones. This lemma entails a prescription which allows to determine the (microscopic) particle density random variables $\mathbf{Q}$ in terms of their macroscopic limits $\underline{\mathcal{Q}}$. In this context it is essential that the scale factor $1/c\kappa$ of the Γ-distributions (4.67), which contains the information on the macroscopic limit, is the same for all $\mathcal{Q}_i$ such that it drops out in the distribution of the proportions (4.72).
(b) For the same reason this lemma can be applied to the Γ-distributions of the continuum limit (eq.(3.145)) only for the the special partition $\mathbf{d} = (1, \ldots, 1)$ corresponding to the occupation numbers. In particular, for BE statistics the exponential distribution (Maxwell-Boltzmann distribution)

$$\Gamma_{\frac{1}{\kappa},1}(x) = \frac{1}{\kappa} \exp(-x/\kappa) \tag{4.73}$$

allows to recover, via the proportions, the uniform de Finetti measure on S_d and, in turn, all formulae of BE statistics.

(c) For necessary conditions under which the Dirichlet distribution is determined via the proportions by Γ-distributions we refer to [121].

(d) By virtue of the deterministic constraint the correlation coefficient of the interchangeable occupation number random variables of Brillouin statistics is determined by the number of cells (see Lemma 3.1.2 and eq.(3.107)). Under the assumptions that the macroscopic particle densities $\underline{Q}$ are independent and that each Q_i is independent of the sum $\sum Q_i$, in [121] it is explained that the negative correlations among the particle densities $\mathbf{Q}$ are correlations which can be thought of due to the constraint $\sum Q_i = 1$ alone.

Finally we consider the 'particle limit', Lemma 3.3.4, from the viewpoint of the de Finetti measure.

Lemma 4.2.5. *Assume that that $d_i \to \infty, \kappa \to 0$ such that $d_i\kappa \to \kappa_i \in \mathbf{R}_+$, then the macroscopic particle densities Q_i converge in distribution to a deterministic observable according to*

$$\Gamma_{\frac{1}{c\kappa},\frac{d_i}{c}}(x) \to \delta_{\kappa_i}(x). \tag{4.74}$$

Proof. Using the characteristic function we obtain

$$\hat{\Gamma}_{\frac{1}{c\kappa},\frac{d_i}{c}}(t) = (1 - c\kappa\, it)^{-d_i/c} \to \{\exp(-it\, c\kappa_i)\}^{-1/c} = \exp(it\, \kappa_i). \tag{4.75}$$

QED

We conclude this section with another characterization of Brillouin statistics. The Pólya-Brillouin distributions for $c > 0$ are mixtures of multinomial distributions. Considering in the multinomial distribution $M_{n,\mathbf{p}}(\mathbf{n})$ the parameters $\mathbf{p}$ as variables and the arguments $\mathbf{n}$ as parameters yields, appropriately renormalized, a continuous probability distribution on $\mathcal{S}_f$, the *conjugate distribution* (see for example [51, 50]). For the multinomial distribution the conjugate distribution is the Dirichlet distribution and in the one-dimensional case the conjugate distribution of the binomial distribution is the beta-distribution (see eq.(4.25)). Accordingly, Brillouin statistics is characterized by the property that the de Finetti measure is the conjugate distribution of the probability distribution which characterizes sums of independent identically distributed random variables with values in $\{1, \ldots, d\}$.

There exist various generalizations of the Pólya process (see for example [97, 91]). These extensions, however, define no interchangeable sequences of random variables. On the other hand, from the viewpoint of statistics with indistinguishable cells and indistinguishable particles there exist obvious generalizations of Brillouin statistics if we use as de Finetti measure a symmetric probability measure on $\mathcal{S}_d$ (see the remark after Theorem 4.2.3).

4.3 Limit Laws of de Finetti's Theorem

In this section we are concerned with limit laws for interchangeable arrays of random variables which generalize the classical Poisson limit theorem and the classical CLT. We use throughout the equivalence between convergence in distribution (denoted by $\xrightarrow{\mathcal{D}}$), convergence of the associated characteristic functions and weak convergence of the induced probability measures.

Interchangeable Triangular Arrays.

Definition 4.3.1. *Let there be given for any $n \in \mathbf{N}, n \geq n_0 \geq 1$, random variables $X_{n,i} : \Omega_n \to \mathbf{R}, 1 \leq i \leq M(n)$, defined on a sequence of probability spaces (Ω_n, F_n, P_n), where $M(n)$ is nondecreasing and $\lim_{n\to\infty} M(n) = \infty$ holds. This triangular array is called an* interchangeable array *if for any fixed n the sequence $X_{n,i}, 1 \leq i \leq M(n)$, is interchangeable. An interchangeable array is called an* ∞-extendible interchangeable array *if the interchangeable random variables in any row are ∞-extendible. E_n denotes the expectation with respect to P_n.*

Remark 4.3.1. The structure of the array is chosen such that it is not necessary to make any assumptions concerning the statistical dependence between observables in different rows.

Any element of $M_+^1(\mathbf{Z}_+)$ can be considered as a limit of an interchangeable array with values in $\{0, 1\}$.

Lemma 4.3.1. *Let there be given a probability measure $\Pi(\cdot)$ on $\mathbf{Z}_+$. Then there exists an interchangeable array $G_{n,i}$ with values in $\{0, 1\}$ such that for any $k \in \mathbf{Z}_+$ we have*

$$\lim_{n\to\infty} P_n(\sum_{i=1}^{n} G_{n,i} = k) = \Pi(k). \tag{4.76}$$

Proof. (i) Assume that $\Pi(0) \neq 0$ and define the interchangeable array by $M(n) = n$ and

$$P_n(G_{n,1} = \varepsilon_1, \ldots, G_{n,n} = \varepsilon_n) = \Pi(k)\{\binom{n}{k} \sum_{i=0}^{n} \Pi(i)\}^{-1} \tag{4.77}$$

where $\varepsilon \in \{0, 1\}^n$ and $k = \sum \varepsilon_i$. Obviously, this defines an interchangeable array and we have

$$P_n(\sum_{i=1}^{n} G_{n,i} = k) = \frac{\Pi(k)}{\sum_{i=0}^{n} \Pi(i)}. \tag{4.78}$$

Since

$$\lim_{n\to\infty} P_n(\sum_{i=1}^{n} G_{n,i} = k) = \Pi(k), \tag{4.79}$$

the interchangeable array has the desired property. (ii) Assume that $0 < n_0 = \min\{i \in \mathbf{Z}_+ ; \Pi(i) \neq 0\}$. For $n \geq n_0$ we define the interchangeable array as in (i). QED

Remark 4.3.2.

(a) Obviously, such an interchangeable array is not uniquely determined. We refer, for example, to the approximation of the Poisson distribution by means of the binomial distribution (independent interchangeable random variables) and by means of the hypergeometric distribution (dependent, but asymptotically independent interchangeable random variables, see Lemma 4.3.7 below). In the limit, however, only the asymptotic properties, which are uniquely determined, are essential.

(b) We emphasize that the construction of some model (array) for a probability measure on $\mathbf{Z}_+$ is necessary for an understanding of the properties of this measure. Without a model it is impossible to infer, for example, that the Poisson distribution describes the number of independent objects as in the limit under consideration the information concerning the correlations of the individual objects is is lost. Lemma 4.3.1 guarantees, for any probability distribution on $\mathbf{Z}_+$, the existence of a model in terms of indistinguishable objects.

(c) A quantum generalization [20] shows in combination with the results in [21] that for any element of $M_+^1(\mathbf{R})$ which is absolutely continuous with respect to the Lebesgue measure an analogous model in terms of an interchangeable array with values in $\{-1, 1\}$ exists.

The following picture explains the structure considered in this section. In this picture dots indicate the structure of the arrays (the lower - centered - array is upside down). Arrows without description indicate the directions considered.

The limit we shall analyze, $\mathcal{X} = \mathcal{X}_0 + \underline{\mathcal{X}}$, is a relation between three random variables that emerge from an ∞-extendible interchangeable array of random variables $X_{n,i}$ in terms of sums. The random variable $\mathcal{X}$ is the limit of the $(n^{-\delta})$-scaled diagonal sums, whereas the random variable $\mathcal{X}_0$ is the limit of the $(n^{-\delta})$-scaled diagonal centered sums. On the other hand, $\underline{\mathcal{X}}$ is the limit of the $(n^{1-\delta})$-scaled mean values. The structure of the array is shown in figure 4.1.

The convergence $n^{1-\delta}\underline{X_n} \xrightarrow{\mathcal{D}} \underline{\mathcal{X}}$ is in general an assumption in our theorems. For the Poisson limit we set $\delta = 0$ whereas for the CLT we set $\delta = 1/2$.

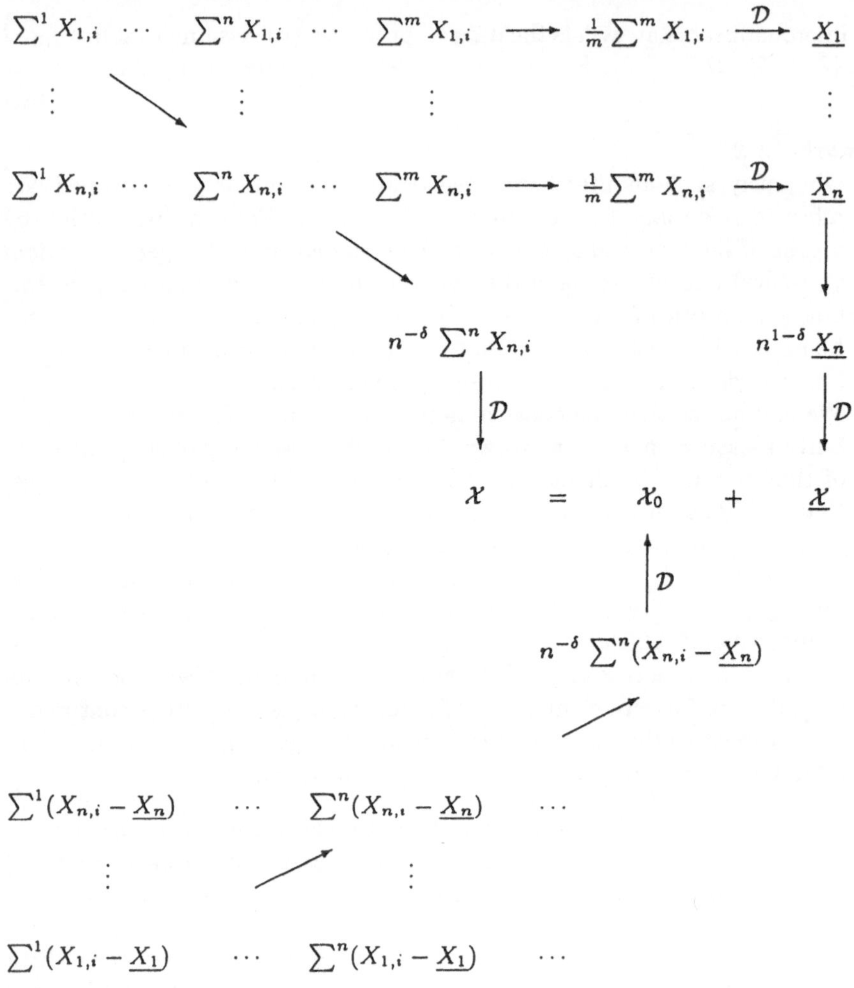

Fig. 4.1. Structure of the ∞–extendible array

4.3.1 The Poisson Limit

Poisson's Theorem.

Theorem 4.3.1. *(Cf. e.g. [40]) Let there be given an ∞–extendible interchangeable array $G_{n,i} : \Omega_n \to \{0,1\}$ and suppose that for any fixed $n \in \mathbf{N}$ the random variables $G_{n,i}$ are independent and identically distributed. For any n set $p_n = P_n(G_{n,i} = 1)$ so that*

$$P_n(\sum_{i=1}^{n} G_{n,1} = k) = B_{n,p_n}(k) \tag{4.80}$$

holds and assume that for $n \to \infty$

$$p_n \to 0 \quad \textit{such that} \quad \lim_{n \to \infty} n\, p_n = x \in \mathbf{R}_+ \tag{4.81}$$

is fulfilled. Then the following equivalent propositions hold.

1. *For all* $k \in \mathbf{Z}_+$

$$\lim_{n \to \infty} P_n(\sum_{i=1}^{n} G_{n,i} = k) = \pi_x(k), \tag{4.82}$$

   where

$$\pi(x) = \frac{1}{k!} e^{-x}\, x^k \tag{4.83}$$

   denotes the Poisson distribution with mean $x \in \mathbf{R}_+$.

2. *For* $t \in \mathbf{R}$

$$\lim_{n \to \infty} E_n \exp(it \sum_{i=1}^{n} G_{n,i}) = \exp\{x\,(e^{it} - 1)\} = \hat{\pi}_x(t). \tag{4.84}$$

Remark 4.3.3.
(a) For a simple quantum generalization we refer to [11].
(b) Assume that $\mathcal{N}$ is distributed according to π_x, then we can define a centered random variable $\mathcal{N}_0 = \mathcal{N} - x$ such that $E\mathcal{N}_0 = 0$.

Benczur's Theorem.

Theorem 4.3.2. *(Cf. [10, 13, 21]) Let there be given, for any* $n \in \mathbf{N}$, *a probability space* (Ω_n, F_n, P_n) *and a sequence of interchangeable random variables* $G_{n,i} : \Omega_n \to \{0,1\}, i \in \mathbf{N}$, *such that*

$$\lim_m \frac{1}{m} \sum_{i=1}^{m} G_{n,i} = Q_n : \Omega_n \to [0,1] \tag{4.85}$$

holds P_n*-a.e. and determines the de Finetti measure* $\nu_n \in M_+^1([0,1])$ *of the sequence* $G_{n,i}, i \in \mathbf{N}$ *by* $\nu_n = P_n^{Q_n}$. *Suppose that* $\mu_n \in M_+^1(\mathbf{R}_+)$, *the image of* ν_n *under the transformation* $t_n : [0,1] \to \mathbf{R}_+, t_n(p) = np$, *converges weakly to* $\mu \in M_+^1(\mathbf{R}_+)$. *Then there exist random variables* $\mathcal{N}, \mathcal{N}_0, \underline{\mathcal{N}}$ *defined on a probability space* (Ω, F, P) *such that for* $n \to \infty$ *the following relations hold.*

1. *Limit of the densities (cf. eq.(4.81)):*

$$Q_n \xrightarrow{\mathcal{D}} 0 \quad \textit{such that} \quad n\, Q_n \xrightarrow{\mathcal{D}} \underline{\mathcal{N}}, \tag{4.86}$$

   where $P^{\underline{\mathcal{N}}} = \mu$.

2. *Centered Poisson limit:*

$$\sum_{i=1}^{n}(G_{n,i} - Q_n) \xrightarrow{\mathcal{D}} \mathcal{N}_0, \tag{4.87}$$

where $\mathcal{N}_0$ is distributed according to a mixture of shifted Poisson distributions,

$$E\exp(it\mathcal{N}_0) = E\exp\{\underline{\mathcal{N}}(e^{it} - 1 - it)\}. \tag{4.88}$$

3. *Poisson limit (Benczur [29]):*

$$\sum_{i=1}^{n} G_{n,i} \xrightarrow{\mathcal{D}} \mathcal{N} \tag{4.89}$$

where

$$E\exp(it\mathcal{N}) = E\exp\{\underline{\mathcal{N}}(e^{it} - 1)\}. \tag{4.90}$$

4. *Structure of the limit:*

$$\mathcal{N} = \mathcal{N}_0 + \underline{\mathcal{N}}, \tag{4.91}$$

where for $(x, y) \in \mathbf{R}^2$

$$E\exp(ix\mathcal{N}_0 + iy\underline{\mathcal{N}}) = E\exp\{\underline{\mathcal{N}}(e^{ix} - 1 - ix + iy)\}. \tag{4.92}$$

Proof. 1. This is a reformulation of the assumption of weak convergence.
2. Using

$$E\{G_{n,1}\,|\,Q_n\} = Q_n, \tag{4.93}$$

we evaluate the characteristic function

$$\begin{aligned} E_n\exp\{it\sum_{i=1}^{n}(G_{n,i} - Q_n)\} &= E_n\Big(\exp(-itn\,Q_n)\,E_n\{\prod^{n}\exp(it\,G_{n,i})\,|\,Q_n\}\Big) \\ &= E_n\Big(\exp(-itn\,Q_n)\Big(E_n\{\exp(it\,G_{n,1})\,|\,Q_n\}\Big)^n\Big) \\ &= E_n\Big(\exp(-itn\,Q_n)\Big(e^{it}Q_n + 1 - Q_n\Big)^n\Big) \\ &= E_n\Big(\exp(-itn\,Q_n)\Big(1 + \frac{n\,Q_n\,(e^{it} - 1)}{n}\Big)^n\Big) \end{aligned} \tag{4.94}$$

By virtue of eq.(4.86) we obtain

$$\begin{aligned} \lim_{n\to\infty} E_n\exp\{it\sum_{i=1}^{n}(G_{n,i} - Q_n)\} &= E\Big(\exp(-it\,\underline{\mathcal{N}})\exp\{\underline{\mathcal{N}}(e^{it} - 1)\}\Big) \\ &= E\exp\{\underline{\mathcal{N}}(e^{it} - 1 - it)\}. \end{aligned} \tag{4.95}$$

3. We have

$$
\begin{aligned}
& E_n \exp(it \sum_{i=1}^{n} G_{n,i}) && (4.96)\\
&= E_n\Big(E_n\{\prod^{n} \exp(it\, G_{n,i}) \,|\, Q_n\}\Big)\\
&= E_n\Big((E_n\{\exp(it\, G_{n,1})|\, Q_n\})^n \Big)\\
&= E_n(e^{it}\, Q_n + 1 - Q_n)^n = E_n(1 + \frac{n\, Q_n\,\{e^{it} - 1\}}{n})^n .
\end{aligned}
$$

By virtue of eq.(4.86) we obtain

$$\lim_n E_n \exp(it \sum_{i=1}^{n} G_{n,i}) = E \exp\{\underline{\mathcal{N}}(e^{it} - 1)\} = E \exp(it\, \mathcal{N}). \quad (4.97)$$

4. From part 2 and 3 we have

$$
\begin{aligned}
E\{\exp(it\, \mathcal{N}_0) \,|\, \underline{\mathcal{N}}\} &= \exp\{\underline{\mathcal{N}}\,(e^{it} - 1 - it)\} && (4.98)\\
E\{\exp(it\, \mathcal{N}) \,|\, \underline{\mathcal{N}}\} &= \exp\{\underline{\mathcal{N}}\,(e^{it} - 1)\}
\end{aligned}
$$

such that $E \exp(it\mathcal{N}) = E \exp(it(\mathcal{N}_0 + \underline{\mathcal{N}}))$ from which $\mathcal{N} = \mathcal{N}_0 + \underline{\mathcal{N}}$ follows. By means of these conditional expectations we obtain

$$
\begin{aligned}
& E \exp(ix\mathcal{N}_0 + iy\underline{\mathcal{N}})\\
&= E\,(E\,\{\exp(ix\mathcal{N}_0 + iy\underline{\mathcal{N}}) \,|\, \underline{\mathcal{N}}\})\\
&= E\,(\exp(iy\underline{\mathcal{N}})\, E\,\{\exp(ix\mathcal{N}_0) \,|\, \underline{\mathcal{N}}\})\\
&= E \exp\{\underline{\mathcal{N}}(e^{ix} - 1 - ix + iy)\}. && (4.99)
\end{aligned}
$$

QED

Remark 4.3.4.
(a) The random number of particles $\mathcal{N}$ is given by a mixture of Poisson distributions, $k \in \mathbf{Z}_+, \mu = P^{\underline{\mathcal{N}}}$,

$$P(\mathcal{N} = k) = \int d\mu(x) \pi_x(k). \quad (4.100)$$

(b) The conditional probability of $\mathcal{N}$, given $\underline{\mathcal{N}}$, is determined by the formula

$$P(\mathcal{N} = k \,|\, \underline{\mathcal{N}}) = \pi_{\underline{\mathcal{N}}}(k). \quad (4.101)$$

Therefore the conditional expectation of $f(\mathcal{N})$ reads

$$E\Big(f(\mathcal{N}) \,|\, \underline{\mathcal{N}}\Big) = \sum_k f(k)\, \pi_{\underline{\mathcal{N}}}(k). \quad (4.102)$$

(c) $\mathcal{N}_0$ and $\underline{\mathcal{N}}$ are statistically dependent iff $\underline{\mathcal{N}}$ is nondeterministic.

(d) In the theorem the assumption of weak convergence $\mu_n \to \mu$ is necessary. Whereas the classical de Finetti theorem, Theorem 4.1.1, includes the weak convergence of the sequence of probability measures, in the Poisson limit, due to lack of compactness, we have to assume weak convergence. From the physical viewpoint weak convergence means that asymptotically the systems in any row of the array are not too different such that the total probability is conserved in the limit.

(e) For the quantum generalization of the Poisson limit of de Finetti's theorem we refer to [21].

Lemma 4.3.2. *For all $n \in \mathbf{Z}_+$ (i)*

$$E(\mathcal{N}^n|\underline{\mathcal{N}}) = \sum_{k=0}^{n} \left\{ \begin{matrix} n \\ k \end{matrix} \right\} \underline{\mathcal{N}}^k, \tag{4.103}$$

where for $0 \leq k \leq n$ the symbol $\left\{ \begin{matrix} n \\ k \end{matrix} \right\}$ denotes the Stirling numbers of the second kind (see e.g. [83]), and (ii) for the descending factorials $\mathcal{N}_{[n]}$

$$E\Big(\mathcal{N}_{[n]}\,|\,\underline{\mathcal{N}}\Big) = \underline{\mathcal{N}}^n. \tag{4.104}$$

Proof. For the evaluation of $E(\mathcal{N}^n|\underline{\mathcal{N}})$ we start from the characteristic function

$$E\big(\exp(it\mathcal{N})|\underline{\mathcal{N}}\big) = E\Big(\exp\left(\underline{\mathcal{N}}\,(e^{it} - 1)\right)\Big). \tag{4.105}$$

The expansion of this expression in terms of powers of it is derived from the formula (cf. [83])

$$\exp\left(x\,(e^t - 1)\right) = \sum_{n=0}^{\infty} \frac{t^n}{n!} \sum_{k=0}^{n} \left\{ \begin{matrix} n \\ k \end{matrix} \right\} x^k, \tag{4.106}$$

and gives (i). We compare this with the expansion of the monomial x^n in terms of falling factorial powers

$$x^n = \sum_{k=0}^{n} \left\{ \begin{matrix} n \\ k \end{matrix} \right\} x_{[k]}, \tag{4.107}$$

and obtain, setting $x = \mathcal{N}$, by evaluation of the conditional expectation

$$E(\mathcal{N}^n|\underline{\mathcal{N}}) = \sum_{k=0}^{n} \left\{ \begin{matrix} n \\ k \end{matrix} \right\} E(\mathcal{N}_{[k]}\,|\,\underline{\mathcal{N}}), \tag{4.108}$$

which, by comparison, yields (ii). QED

Remark 4.3.5. Using the results of [21] we obtain the following quantum identities related to the transformation to normal order (cf. e.g. [102] for part (ii)).

On the domain given by the finite linear combinations of eigenvectors $\Phi(n), n \in \mathbf{Z}_+$, of the number operator a^+a or evaluated by the linear functionals $E_{z,z'}, z, z' \in \mathbf{C}$,

$$E_{z,z'}(\cdot) = <\Psi(z), \cdot > \Psi(z'), \tag{4.109}$$

where $\Psi(z), z \in \mathbf{C}$, is a coherent state vector, for all $n \in \mathbf{Z}_+$

(i)

$$(a^+a)^n = \sum_{k=0}^{n} \left\{ \begin{matrix} n \\ k \end{matrix} \right\} a^{+k} a^k, \tag{4.110}$$

(ii)

$$(a^+a)_{[n]} = a^{+n} a^n. \tag{4.111}$$

Examples.

Example 4.3.1. Brillouin Statistics.

In any row of the interchangeable array we assume that the random variables $G_{n,i}$ are distributed according to Brillouin statistics ($c > 0, f = 2, d_1 = 1, n_1 = k, d = d(n)$). According to Theorem 4.1.1 and eq.(4.25), any row of the array is characterized by a normalized particle density Q_n distributed according to a beta distribution

$$b_{\frac{d(n)-1}{c}, \frac{1}{c}}(p) = \frac{\Gamma(\frac{d(n)}{c})}{\Gamma(\frac{d(n)-1}{c})\,\Gamma(\frac{1}{c})} p^{(\frac{1}{c}-1)}(1-p)^{(\frac{d(n)-1}{c}-1)}. \tag{4.112}$$

We now assume that $d(n) \to \infty$ for $n \to \infty$ such that $n/d(n) \to \overline{n}$ holds (macroscopic limit). In this limit we obtain

$$E_n \exp(it\, n\, Q_n) = \int dx \exp(itx) \frac{\Gamma(\frac{d(n)}{c})}{\Gamma(\frac{1}{c})\,\Gamma(\frac{d(n)-1}{c})} \frac{x^{(\frac{1}{c}-1)}}{n^{\frac{1}{c}}} \{1 - \frac{x}{n}\}^{(\frac{d(n)-1}{c}-1)}, \tag{4.113}$$

so that in the limit the de Finetti measure is a $\Gamma_{\frac{1}{c\overline{n}}, \frac{1}{c}}$-distribution

$$\begin{aligned} \hat{\mu}(t) &= E \exp(it\, \underline{\mathcal{N}}) = \lim_{n\to\infty} E \exp(it\, n\, Q_n) \\ &= \int dx \exp(itx) \frac{(\frac{1}{c\overline{n}})^{\frac{1}{c}}}{\Gamma(1/c)} x^{(\frac{1}{c}-1)} \exp(-\frac{x}{c\overline{n}}). \end{aligned} \tag{4.114}$$

In particular, we obtain the integral representation

$$\begin{aligned} P(\mathcal{N} = k) &= \int dx\, \Gamma_{\frac{1}{c\overline{n}}, \frac{1}{c}}(x)\, \pi_x(k) \\ &= \binom{\frac{1}{c} + k - 1}{k} (\frac{1}{1 + c\overline{n}})^{\frac{1}{c}} (\frac{c\overline{n}}{1 + c\overline{n}})^k, \end{aligned} \tag{4.115}$$

for the negative binomial distribution in terms of Poisson distributions (cf. [63, p.57]). These probability distributions have been derived in Section 3.3, eq.(3.131), by means of the macroscopic limit such that we conclude that $P(\mathcal{M} = k) = P(\mathcal{N} = k), k \in \mathbf{Z}_+$, holds. Moreover, the Γ-distributions which act as representing measures have been determined in Section 3.3 as macroscopic continuum distributions. Finally, the same Γ-distributions have been determined in Section 4.2 as the macroscopic limits of the Dirichlet distribution. Accordingly, for Brillouin statistics ($c \geq 0$) the macroscopic limit of the de Finetti measure is equal to the de Finetti measure of the macroscopic limit.

Mixtures and Convolutions. Limit laws of de Finetti's theorem constitute mixtures of Poisson distributions,

$$\begin{aligned}\mathcal{M}_\pi &= \{\Pi \in M_+^1(\mathbf{Z}_+);\ \Pi(k) = \\ &= \int d\mu(x)\pi_x(k)\,, k \in \mathbf{Z}_+, \text{ for some } \mu \in M_+^1(\mathbf{R}_+)\}.\end{aligned} \tag{4.116}$$

Lemma 4.3.3. *(Cf. [22].) The convex set of mixtures of Poisson distributions $\mathcal{M}_\pi$ is a simplex which is generated in the weak topology of $M_+^1(\mathbf{Z}_+)$ as the closed convex hull of its extreme points*

$$\mathcal{M}_\pi = \mathrm{cl}\,(\mathrm{con}\,(\mathrm{ex}\,(\mathcal{M}_\pi))), \tag{4.117}$$

where the extreme points are the Poisson distributions

$$\mathrm{ex}\,(\mathcal{M}_\pi) = \{\pi_x \in M_+^1(\mathbf{Z}_+);\ x \in \mathbf{R}_+\}. \tag{4.118}$$

Remark 4.3.6.
(a) The set $\mathcal{M}_\pi$ is not compact in the weak topology such that the simplex under consideration is no Choquet simplex.
(b) As any element of a simplex admits a unique barycentric decomposition in terms of extreme elements this lemma guarantees the uniqueness of the de Finetti measures in Theorem 4.3.2.
(c) The probability distributions that are analyzed in this section have the structure

$$P(\mathcal{N} = k) = \int d\mu(x)\,\pi_x(k) \tag{4.119}$$

where $\mu = P^{\underline{\mathcal{N}}} \in M_+^1(\mathbf{R}_+)$ and $\underline{\mathcal{N}}$ is the macroscopic particle density. The expectation and the variance of $\mathcal{N}$ are given by the formulae

$$E\mathcal{N} = E\underline{\mathcal{N}}, \tag{4.120}$$

$$\mathrm{Var}(\mathcal{N}) = E\underline{\mathcal{N}} + \mathrm{Var}(\underline{\mathcal{N}}). \tag{4.121}$$

Therefore, in general the variance consists of two non-negative terms, corresponding to the variance and expectation of the macroscopic particle density, respectively (this is the structure that is imposed by the Poisson distribution).

(d) For any fixed value of $E\mathcal{N} = E\underline{\mathcal{N}}$ the variance of $\mathcal{N}$ is minimal if $P^{\underline{\mathcal{N}}} \in \mathrm{ex}\,(\mathcal{M}_\pi)$, that is, if the de Finetti measure is a Dirac measure. Obviously, mixtures of Poisson distributions belong to the class of super-Poisson statistics.

(e) The structure of the mixtures determined by Theorem 4.3.2 entails the following expression for the variance

$$\mathrm{Var}(\mathcal{N}) = E\,\{\mathrm{Var}(\mathcal{N}|\underline{\mathcal{N}})\,\} + \mathrm{Var}\,\{\,E(\mathcal{N}|\underline{\mathcal{N}})\,\} \tag{4.122}$$

where the conditional variance is defined by

$$\mathrm{Var}(\mathcal{N}|\underline{\mathcal{N}}) = E\,(\mathcal{N}^2\,|\,\underline{\mathcal{N}}) - E^2\,(\mathcal{N}\,|\,\underline{\mathcal{N}}). \tag{4.123}$$

(f) The Poisson distribution, characterizing independent particles, is determined by a Dirac measure as de Finetti measure. Accordingly, $\mathrm{Var}(\underline{\mathcal{N}}) = 0$ holds iff only the first term in eq.(4.121) contributes. Therefore, the second term of $\mathrm{Var}(\mathcal{N})$ represents precisely that contribution that is due to an asymptotic statistical dependence among the particles.

It is well-known that Poisson distributions are closed under convolutions (see e.g. [62])

$$\pi_x \star \pi_y = \pi_{x+y}. \tag{4.124}$$

This property extends to mixtures where in the following Π_μ denotes a generic element of the simplex $\mathcal{M}_\pi$ with de Finetti measure $\mu \in M^1_+(\mathbf{R}_+)$.

Lemma 4.3.4. *Mixtures of Poisson distributions are closed under convolutions and the de Finetti measure of a convolution is the convolution of the de Finetti measures*

$$\Pi_\mu \star \Pi_\nu = \Pi_{\mu\star\nu} \tag{4.125}$$

Proof. According to the multiplicative structure of characteristic functions under convolutions we have

$$\begin{aligned}
\widehat{\Pi_\mu \star \Pi_\nu}(t) &= \hat{\Pi}_\mu(t)\,\hat{\Pi}_\nu(t)\\
&= \int d\mu(x)\,\exp(x\,(e^{it}-1))\int d\nu(y)\,\exp(y\,(e^{it}-1))\\
&= \int\!\!\int d\mu(x)\,d\nu(y)\exp((x+y)\,(e^{it}-1))\\
&= \int d(\mu\star\nu)(z)\exp(z\,(e^{it}-1)).
\end{aligned} \tag{4.126}$$

QED

Remark 4.3.7.

(a) Mixtures of Poisson distributions have been studied extensively in the context of photoelectron statistics (see e.g. [135]) where the mapping $\mathcal{P} : M_+^1(\mathbf{R}_+) \to M_+^1(\mathbf{Z}_+)$, defined by

$$\mathcal{P}(\mu) = \Pi_\mu \tag{4.127}$$

is called *Poisson transformation.* In the theory of photoelectron statistics the de Finetti measure is the distribution of the integrated density of photoelectrons and in this context the determination of μ from the number of photoelectron counts, given by $\Pi \in M_+^1(\mathbf{Z}_+)$, is the interesting quantity.

(b) The methods proposed so far to determine the mixing measure (see for example [135]) treat mixtures of Poisson distributions as abstract structures, that is, without any connection to the characterization of such mixtures by means of the Poisson limit of de Finetti's theorem. Obviously, the method based on the expansion of a de Finetti measure μ in a series of Laguerre polynomials [135] (also the basis of the extendibility investigations in [96]) is restricted to those μ that have a density with respect to the Lebesgue measure and fulfil the necessary integrability conditions.

In our pre-eminent example (Brillouin statistics) the macroscopic particle density $\underline{\mathcal{N}}$ is distributed according to a $\Gamma_{1/c\overline{n},d_1/c}$-distribution (see eq.(4.113) for $d_1 = 1$) and it is well-known that Γ-distributions are closed under convolutions if the scale factor α is fixed

$$\Gamma_{\alpha,\beta} \star \Gamma_{\alpha,\gamma} = \Gamma_{\alpha,\beta+\gamma} \tag{4.128}$$

(see e.g. [63]). The macroscopic particle density $\underline{\mathcal{N}}$ therefore can be considered as the sum of independent identically distributed random variables $\underline{\mathcal{N}}_i, 1 \leq i \leq d_1$, where $\underline{\mathcal{N}}_1$ is distributed according to a $\Gamma_{1/c\overline{n},1/c}$-distribution. On the level of the de Finetti measures this property reflects precisely that the negative binomial distribution

$$n_{c\overline{n},d_1/c}(k) = \begin{pmatrix} \frac{d_1}{c} + k - 1 \\ k \end{pmatrix} \left(\frac{1}{1+c\overline{n}}\right)^{d_1/c} \left(\frac{c\overline{n}}{1+c\overline{n}}\right)^k \tag{4.129}$$

is a convolution

$$n_{c\overline{n},d_1/c} = \overset{d_1}{\star}\, n_{c\overline{n},1/c} \tag{4.130}$$

(see e.g. [62]).

Remark 4.3.8.

(a) We have characterized elements of the simplex $\mathcal{M}_\pi$ by means of the Poisson limit of de Finetti's theorem, that is, by means of the concept of indistinguishability. It is well known that for some elements of this set also another model is conceivable in terms of *Poisson sums* (the distribution is called a *compound Poisson distribution* in [62]). For the equivalence of Poisson sums to mixtures of Poisson distributions we refer to [81].

(b) At the end of Section 4.2 we noticed that the beta–distribution is the conjugate distribution of the binomial distribution. On the level of the Poisson limit of de Finetti's theorem this correspondence still exists. The conjugate distribution of the Poisson distribution $\pi_x(k)$ is the $\Gamma_{1,k+1}(x)$-distribution.
(c) For the quantum generalization of Lemma 4.3.3 we refer to [23].
(d) For the quantum generalization of the convolution properties we refer to [43].

In the context of the integral representations conjugate distributions play an important role. From the viewpoint of quantum theory they are connected with the existence of a continuous resolution of the identity on the underlying Hilbert space in terms of discrete coherent states or coherent states, respectively (cf. Lemma 2.3.8).

In this context there exists an interrelation between generalizations of de Finetti's theorem and the theory of generalized coherent states. We are concerned with integral representations in terms of the binomial distribution and its natural generalizations, the multinomial distribution and the Poisson distribution. The quantum theory of general coherent states [129] associates with any Lie group a total set of generalized coherent vectors and the associated generalized coherent states. An integral representation of states in terms of these generalized coherent states defines, in appropriate abelian subalgebras of observables, a mixture of probability distributions. For mixtures (with respect to the mean value) of negative binomial distributions and their connection with the hydrogen atom we refer the reader to [77].

Problem 4.3.1. Are the general coherent states in the sense of PERELOMOV [129] in general MB symmetric or limits of MB symmetric states?

Galambos's Theorem. We now no longer assume that the random variables in any row of the interchangeable array are ∞–extendible, but are interested in the necessary extendibility properties of the rows of an interchangeable array for the convergence to a mixture of Poisson distributions. We consider an interchangeable array $G_{n,i} : \Omega \to \{0,1\}$, where $1 \le i \le M(n)$ and assume that the sequence $G_{n,i}$ is, for any fixed n, $N(n)$–extendible, $N(n) > M(n)$. According to de Finetti's formula, eq.(4.1), we have for any $n \in \mathbf{N}$ and any $k, 0 \le k \le M(n)$,

$$P_n(\sum_{i=1}^{M(n)} G_{n,i} = k) = \sum_{r=k}^{N(n)-M(n)+k} P_n(\sum_{i=1}^{N(n)} G_{n,i} = r)\ \mathcal{H}_{N(n),M(n),r}(k). \tag{4.131}$$

Introducing the random variables

$$\mathcal{R}(n) = \frac{M(n)}{N(n)} \sum_{i=1}^{N(n)} G_{n,i} \tag{4.132}$$

and the associated induced probability measures $\mu_n \in M_+^1(\mathbf{R}_+), \mu_n = P_n^{\mathcal{R}(n)}$, eq.(4.131) can be rewritten as

$$P_n(\sum_{i=1}^{M(n)} G_{n,i} = k) = \int_{\alpha(n)}^{\beta(n)} d\mu_n(x)\, \mathcal{H}_{N(n),M(n),N(n)x/M(n)}(k), \qquad (4.133)$$

where

$$\alpha(n) = \frac{M(n)k}{N(n)} \quad \text{and} \quad \beta(n) = \frac{M(n)\{N(n) - M(n) + k\}}{N(n)}. \qquad (4.134)$$

To motivate the strategy which is used to derive an integral representation from this formula we first give another direct proof of Benczur's theorem (Theorem 4.3.2 part 3) based upon a general lemma concerning the limit of an expectation when both the measure and the random variable vary.

Lemma 4.3.5. *(Cf. [24]) Assume that the sequence of probability measures* $\mu_n \in M_+^1(\mathbf{R}_+)$ *converges weakly to a probability measure* $\mu \in M_+^1(\mathbf{R}_+)$. *Moreover, suppose that the sequence of functions* $f_n \in C_b(\mathbf{R}_+)$ *is uniformly bounded and converges uniformly on compact sets to a function* $f \in C_b(\mathbf{R}_+)$, *then*

$$\lim_{n\to\infty} \int d\mu_n(x)\, f_n(x) = \int d\mu(x)\, f(x). \qquad (4.135)$$

Lemma 4.3.6. *(Cf. e.g. [62].) For* $p_n = x/n, x \in \mathbf{R}_+$, *for all* $k \in \mathbf{Z}_+$, *the convergence in eq.(4.82) is uniform on compact sets, so that for all* $c \in \mathbf{R}_+$

$$\lim_{n\to\infty} \sup_{x\,\in\,[0,c]} |B_{n,x/n}(k) - \pi_x(k)| = 0. \qquad (4.136)$$

A combination of these lemmata allows us to prove Benczur's theorem in the following form.

Proof. According to the assumptions of an ∞–extendible interchangeable array $G_{n,i}$ we have for any $n \in \mathbf{N}$ and any $k, 0 \le k \le n$,

$$P_n(\sum_{i=1}^{n} G_{n,i} = k) = \int d\nu_n(p) B_{n,p}(k) = \int d\mu_n(x) B_{n,x/n}(k). \qquad (4.137)$$

Therefore, for any k,

$$\begin{aligned} &|\int d\mu(x)\, \pi_x(k) - \int d\mu_n(x)\, B_{n,x/n}(k)| \\ \le\ & |\int d\mu(x)\, \pi_x(k) - \int d\mu_n(x)\, \pi_x(k)| \\ & + |\int d\mu_n(x)\, \pi_x(k) - \int d\mu_n(x)\, B_{n,x/n}(k)| \\ =\ & I_n(k) + II_n(k). \end{aligned} \qquad (4.138)$$

Since $\pi_x(k) \in C_b(\mathbf{R}_+)$ and $\mu_n \to \mu$ weakly, the first contribution $I_n(k)$ vanishes in the limit. Moreover

$$II_n(k) \leq \int d\mu_n(x)|\pi_x(k) - B_{n,x/n}(k)| \tag{4.139}$$

such that $II_n(k)$ vanishes by virtue of Lemma 4.3.6. QED

This proof is based on the Poisson approximation of the binomial distribution. If the array under consideration is not ∞–extendible we need a result on the Poisson approximation of the hypergeometric distribution.

Lemma 4.3.7. *(Cf. e.g. [72, Lemma 3.4.2.]) For all $k \in \mathbf{Z}_+$ the hypergeometric distribution*

$$\mathcal{H}_{n,m,r}(k) \in \binom{n}{r}^{-1} \binom{m}{k} \binom{n-m}{r-k} \tag{4.140}$$

converges in the limit $n, m, r \to \infty$ such that $mr/n \to x > 0$ uniformly on compact sets to the Poisson distribution π_x, so that for any $c \in \mathbf{R}_+$

$$\lim_{n\to\infty} \sup_{x\in[0,c]} |\mathcal{H}_{n,m,r}(k) - \pi_x(k)| = 0. \tag{4.141}$$

Theorem 4.3.3. *(Cf. [70, 72, 71]). Let there be given an interchangeable array $G_{n,i}, n \in \mathbf{N}, 1 \leq i \leq M(n)$, and assume that asymptotically for any n the sequence $G_{n,i}, 1 \leq i \leq M(n)$, is $N(n)$-extendible such that for $n \to \infty$*

$$M(n)/N(n) \to 0 \tag{4.142}$$

is satisfied. Then for all $k \in \mathbf{Z}_+$

$$\lim_{n\to\infty} P_n(\sum_{i=1}^{M(n)} G_{n,i} = k) = \int d\mu(x)\pi_x(k) \tag{4.143}$$

holds iff the sequence of induced probability measures $\mu_n = P_n^{\mathcal{R}(n)}$, converges weakly to a probability measure $\mu \in M_+^1(\mathbf{R}_+)$.

Proof. (Cf. [70, 72, 71]) In this proof the explicit dependence of N and M upon n is suppressed for notational convenience. For sufficiency we start from eq.(4.133) and use Lemma 4.3.5 where we set

$$f_n(x) = \mathcal{H}_{N,M,Nx/M}(k), \qquad f(x) = \pi_x(k). \tag{4.144}$$

According to Lemma 4.3.7 the assumptions of Lemma 4.3.5 are fulfilled such that the assertion follows. For necessity assume that

$$\lim_{n\to\infty} P_n(\sum_{i=1}^{M} G_{n,i} = k) = \int d\mu(x)\pi_x(k) \tag{4.145}$$

holds for all $k \in \mathbf{Z}_+$. According to eq.(4.133)

$$P_n(\sum_{i=1}^{M} G_{n,i} = k) = \int_{\alpha(n)}^{\beta(n)} d\mu_n(x) \mathcal{H}_{M,N,Nx/M}(k) \tag{4.146}$$

holds by construction. Due to the compactness of bounded measures we can select a subsequence $\mu_{n_j}, j \in \mathbf{N}$, which converges to a bounded measure $\tilde{\mu}$. Obviously $\tilde{\mu}$ is a probability measure. Replicating part (i), we obtain a representation in terms of mixtures of Poisson distributions by means of $\tilde{\mu}$. If the sequence of probability measures did not converge weakly, we could choose two subsequences which would converge weakly to two different probability measures. This, however, is impossible since the de Finetti measure is uniquely determined (see Lemma 4.3.3). QED

Remark 4.3.9.
(a) The de Finetti measure is determined by the following characteristic function

$$\hat{\mu}(t) = \lim_{n\to\infty} E_n \exp(it \frac{M(n)}{N(n)} \sum_{i=1}^{N(n)} G_{n,i}). \tag{4.147}$$

(b) Under the extendibility condition (4.142) this theorem states that weak convergence of the sequence μ_n is necessary and sufficient.

In the examples we assume that the following conditions are satisfied where d_n is the number cells in the nth row of the array.

1. $M(n)/N(n) \to 0$ for $n \to \infty$,
2. $d(n) = d_n \to \infty$ such that $M(n)/d_n \to \overline{n}$ holds.

Example 4.3.2. MB Statistics.
We set $P_n(G_{n,1} = 1) = 1/d_n$ and obtain

$$E_n \exp(it \frac{M(n)}{N(n)} \sum_{i=1}^{N(n)} G_{n,i}) = [1 + \frac{1}{d_n} \{\exp(it \frac{M(n)}{N(n)}) - 1\}]^{N(n)}$$
$$= \{1 + \frac{it}{N(n)} \frac{M(n)}{d_n} + o(\frac{1}{N(n)})\}^{N(n)} \to \exp(it\,\overline{n}), \tag{4.148}$$

which is the characteristic function of the Dirac measure concentrated at $\overline{n}$.

Example 4.3.3. Hypergeometric Distribution.
We assume that the random variables of the interchangeable array $G_{n,i}$, $1 \leq i \leq M(n)$, are distributed according to the law

$$P_n(\sum_{i=1}^{M(n)} G_{n,i} = k) = \mathcal{H}_{N(n),M(n),R(n)}(k). \tag{4.149}$$

For any fixed n this sequence is $\max(N(n))$–extendible. In the limit $N(n)$, $M(n)$,
$R(n) \to \infty$ such that (i) $M(n)R(n)/N(n) \to x \in \mathbf{R}_+$ and (ii) $M(n)/N(n) \to 0$ we have according to Lemma 4.3.7

$$\lim_{n\to\infty} P_n(\sum_{i=1}^{M(n)} G_{n,i} = k) = \pi_x(k) \tag{4.150}$$

so that the de Finetti measure is a Dirac measure.

Example 4.3.4. BE Statistics.
Under assumptions (1), (2) above we evaluate

$$\begin{aligned} P_n(\frac{M}{N}\sum_{i=1}^{N} G_{n,i} < x) &= \sum_{r<Nx/M} \binom{d_n+N-1}{N}^{-1} \binom{d_n+N-r-2}{N-r} \\ &= \sum_{r<Nx/M} (d_n-1)\frac{N_{[r]}}{(d_n+N-1)_{[r+1]}}. \end{aligned} \tag{4.151}$$

Up to terms of order $O(\frac{1}{N})$ and $O(\frac{1}{d_n+N})$ we obtain

$$\begin{aligned} &P_n(\frac{M}{N}\sum_{i=1}^{N} Y_{n,i} < x) \\ &= \sum_{r<Nx/M} \frac{d_n-1}{d_n+N-1}(\frac{N}{d_n+N-1})^r = 1 - \{\frac{\frac{N}{d_n}}{1+\frac{N}{d_n}-\frac{1}{d_n}}\}^{Nx/M} \\ &= 1 - \{1+\frac{M}{N}(\frac{d_n}{M}+\frac{1}{M})\}^{-Nx/M} \to 1-\exp(-x/\overline{n}), \end{aligned} \tag{4.152}$$

which is the distribution function of the exponential distribution.

Condition (4.142) guarantees the convergence to a mixture. On the other hand examples show that for $M(n)/N(n) \to p > 0$ we obtain in general no mixture.

Theorem 4.3.4. *Let there be given an interchangeable array $G_{n,i}, n \in \mathbf{N}, 1 \le i \le M(n)$, and assume that the sequence of induced probability measures $\mu_n = P_n^{\mathcal{R}(n)}$, converges weakly to a probability measure $\mu \in M_+^1(\mathbf{R}_+)$. Then for all $k \in \mathbf{Z}_+$*

$$\lim_{n\to\infty} P_n(\sum_{i=1}^{M(n)} G_{n,i} = k) = \int d\mu(x)\pi_x(k) \tag{4.153}$$

holds iff asymptotically for any n the sequence $G_{n,i}, 1 \le i \le M(n)$, is $N(n)$-extendible such that for $n \to \infty$

$$M(n)/N(n) \to 0 \tag{4.154}$$

is satisfied.

Proof. Sufficiency follows from Theorem 4.3.3. That condition (4.142) is necessary is implied be the following examples where the interchangeable array $G_{n,i}, 1 \leq i \leq M(n)$, is asymptotically linear extendible, that is, for fixed n the sequence $G_{n,i}$ is $\max(N(n))$–extendible such that

$$M(n)/N(n) \to p \in (0,1] \tag{4.155}$$

holds. In this case, provided the array converges to an element $\Pi \in M_+^1(\mathbf{Z}_+)$, Π is in general no mixture of Poisson distributions. QED

We give two examples of probability distributions which admit no integral representation in terms of mixtures of Poisson distributions.

Example 4.3.5. Number states [14].

We set $M(n) = n$, fix $k' \in \mathbf{N}$ and define for $n > k', \varepsilon_i \in \{0,1\}, 1 \leq i \leq n$, the interchangeable array $G_{n,i} : \Omega \to \{0,1\}, n \in \mathbf{N}, n > k', 1 \leq i \leq n$, by

$$P_n(G_{n,1} = \varepsilon_1, \ldots, G_{n,n} = \varepsilon_n) = \begin{pmatrix} n \\ k \end{pmatrix}^{-1} \delta_{k'}(k) \tag{4.156}$$

where $\sum_{i=1}^{n} \varepsilon_i = k$. The probability of the sum of these random variables is a Dirac measure

$$P_n(\sum_{i=1}^{n} G_{n,i} = k) = \delta_{k'}(k), \tag{4.157}$$

so that for $k \in \mathbf{N}$

$$\lim_{n\to\infty} P_n(\sum_{i=1}^{n} G_{n,i} = k) = \delta_{k'}(k). \tag{4.158}$$

On the other hand it follows from Lemma 3.2.1 that the sequence $G_{n,i}, 1 \leq i \leq n$, is, for fixed n, not $(n+1)$–extendible. Therefore we have $n = M(n) = N(n)$. That an integral representation is impossible follows from to the well-known fact that the number states of the quantum harmonic oscillator are no classical states (cf. [20]).

Example 4.3.6. Squeezed Vacuum States [14].

We set $M(n) = 2n$ and define the interchangeable random variables $n \in \mathbf{N}, \varepsilon_i \in \{0,1\}, 1 \leq i \leq 2n$,

$$P_n(G_{n,i} = \varepsilon_i, 1 \leq i \leq 2n) = \begin{cases} 0 & \text{if } \sum_{i=1}^{2n} \varepsilon_i \text{ is odd,} \\ \begin{pmatrix} n \\ 2k \end{pmatrix}^{-1} P_n(K_n = k) & \text{if } \sum_{i=1}^{2n} \varepsilon_i = 2k. \end{cases} \tag{4.159}$$

Here for any $n \in \mathbf{N}$ the random variable $K_n : \Omega \to \{0,1,\ldots,n\}$ is distributed according to a Pólya distribution where $d = d_n \in \mathbf{N}, d \geq 2, c > 0$,

$$P_n(K_n = k) = \binom{\frac{1}{c}+n-1}{n}^{-1} \binom{\frac{1}{c}+k-1}{k} \binom{\frac{d-1}{c}+n-k-1}{n-k}. \tag{4.160}$$

We assume that $d_n \to \infty$ for $n \to \infty$ such that $n/d_n \to \overline{n} \in \mathbf{R}_+$. Under these conditions eq.(4.160) converges to a negative binomial distribution and we obtain

$$\lim_{n\to\infty} E_n \exp\{it \sum_{i=1}^{2n} G_{n,i}\} = \{1 + c\overline{n} - c\overline{n} \exp(2it)\}^{-1/c}. \tag{4.161}$$

This shows that a limit distribution exists. According to Lemma 3.2.1 the sequence $G_{n,i}, 1 \le i \le 2n$, is not $(2n+1)$–extendible. For $c = 2$, eq.(4.161) is the characteristic function of the distribution of the number of photons in the squeezed vacuum state (see e.g. [155]). That an integral representation in terms of mixtures of Poisson distributions is not possible follows from the well-known fact that squeezed states are no classical states and admit, therefore, no integral representation in terms of coherent states (cf. [2, 20]).

For an investigation of the asymptotic behaviour of the correlation structure of an unextendible interchangeable array with values in $\{0,1\}$ we refer to [13].

4.3.2 The Central Limit

In this section we are no longer concerned with the statistics of particles (or the numbers of indistinguishable objects) but with indistinguishable objects with values in $\{-1,1\}$ which may be identified with elementary fields.

CLT for Sequences. Our first goal is the generalization of the familiar CLT for i.i.d. random variables to interchangeable random variables. To this end we recall that for an interchangeable sequence $X_i, i \in \mathbf{N}, E|X_1| < \infty$, the strong LLN defines a random variable $\underline{X}$ which induces the representing measure. Moreover, by virtue of conditional identity we have for any $n \in \mathbf{N}$

$$E\{X_1 \,|\, \underline{X}\} = E\{\frac{1}{n}\sum_{i=1}^{n} X_i \,|\, \underline{X}\}, \tag{4.162}$$

such that, by virtue of the strong LLN,

$$E\{X_1 \,|\, \underline{X}\} = \underline{X} \tag{4.163}$$

holds.

Theorem 4.3.5. *(Cf. e.g. [4, 40]) Let there be given interchangeable random variables $X_i : (\Omega, F, P) \to \mathbf{R}, i \in \mathbf{N}$, and suppose that $E|X_1| < \infty$, such that*

$$\lim_n \frac{1}{n} \sum_{i=1}^{n} X_i = \underline{X} \tag{4.164}$$

exists P-a.e., then, if $E|X_1|^2 < \infty$,

$$\frac{1}{\sqrt{n}} \sum_{i=1}^{n} \{X_i - \underline{X}\} \overset{\mathcal{D}}{\to} \Sigma(\underline{X})\, X_0, \tag{4.165}$$

where

$$\Sigma^2(\underline{X}) = E\{(X_1 - \underline{X})^2 | \underline{X}\} \tag{4.166}$$

is the conditional variance and X_0 *is independent from* $\underline{X}$ *and distributed according to* $N(0,1)$.

Proof. The random variables X_i are i.i.d. conditionally on $\underline{X}$. Therefore,

$$\begin{aligned}
&\lim_n E \exp\{it \frac{1}{\sqrt{n}} \sum_{i=1}^{n} (X_i - \underline{X})\} \\
&= \lim_n E\{ E(\exp[it \frac{1}{\sqrt{n}} \sum_{i=1}^{n} (X_i - \underline{X})] \,|\, \underline{X})\} \\
&= \lim_n E\{ E(\exp[it \frac{1}{\sqrt{n}} (X_1 - \underline{X})] \,|\, \underline{X})\}^n \\
&= \lim_n E\Big(1 - \frac{t^2}{2n} E\{(X_1 - \underline{X})^2 \,|\, \underline{X}\} + o(\frac{1}{n})\Big)^n \\
&= E \exp\{-\frac{(t\Sigma(\underline{X}))^2}{2}\} = E \exp\{it\, X_0\, \Sigma(\underline{X})\}.
\end{aligned} \tag{4.167}$$

QED

Remark 4.3.10. Accordingly, we obtain mixtures of centered Gaussian distributions, where the mixing concerns the variance (cf. [63]), $\nu \in M_+^1(\mathbf{R}_+)$,

$$p(x) = \int d\nu(\sigma) \frac{1}{\sqrt{2\pi\sigma^2}} \exp(-\frac{x^2}{2\sigma^2}). \tag{4.168}$$

Example 4.3.7. Assume that the random variables $X_i : \Omega \to \{-1, 1\}, i \in \mathbf{N}$, are interchangeable and denote the de Finetti measure by $\nu = P^{\underline{X}} \in M_+^1([-1,1])$ such that

$$P(X_1 = \pm 1 \,|\, \underline{X}) = \frac{1 \pm \underline{X}}{2}. \tag{4.169}$$

We obtain

$$E\{X_1 \,|\, \underline{X}\} = \underline{X}, \tag{4.170}$$
$$E\{X_1^2 \,|\, \underline{X}\} = 1, \tag{4.171}$$
$$\Sigma^2(\underline{X}) = 1 - \underline{X}^2, \tag{4.172}$$

such that the characteristic function of the limit reads

$$\phi(t) = E \exp\{-\frac{t^2}{2}(1 - \underline{X}^2)\}. \tag{4.173}$$

CLT for Arrays.

Theorem 4.3.6. *(Cf. [21]) Let there be given, for any $n \in \mathbf{N}$, a probability space (Ω_n, F_n, P_n) and a sequence of interchangeable random variables $Y_{n,i} : \Omega_n \to \{-1,1\}, i \in \mathbf{N}$, such that the relation*

$$\lim_m \frac{1}{m} \sum_{i=1}^{m} Y_{n,i} = \underline{Y_n} : \Omega_n \to [-1,1] \tag{4.174}$$

holds P_n-a.e. and determines the de Finetti measure $\nu_n \in M_+^1([-1,1])$ of the sequence $Y_{n,i}, i \in \mathbf{N}$, by virtue of $\nu_n = P_n^{\underline{Y_n}}$. Suppose that $\mu_n \in M_+^1(\mathbf{R})$, the image of ν_n under the transformation $\tau_n : [-1,1] \to \mathbf{R}, \tau_n(x) = \sqrt{n}x$, converges weakly to $\mu \in M_+^1(\mathbf{R})$. Then there exist random variables $\mathcal{F}, \mathcal{F}_0, \underline{\mathcal{F}}$ defined on a probability space (Ω, F, P) such that for $n \to \infty$ the following properties hold.

1. *Limit of the densities:*

$$\underline{Y_n} \overset{\mathcal{D}}{\to} 0 \quad \textit{such that} \quad \sqrt{n}\,\underline{Y_n} \overset{\mathcal{D}}{\to} \underline{\mathcal{F}}, \tag{4.175}$$

   where $P^{\underline{\mathcal{F}}} = \mu$.

2. *CLT:*

$$\frac{1}{\sqrt{n}} \sum_{i=1}^{n} (Y_{n,i} - \underline{Y_n}) \overset{\mathcal{D}}{\to} \mathcal{F}_0, \tag{4.176}$$

   where $\mathcal{F}_0$ is distributed according to $N(1,0)$.

3. *Generalized CLT:*

$$\frac{1}{\sqrt{n}} \sum_{i=1}^{n} Y_{n,i} \overset{\mathcal{D}}{\to} \mathcal{F}, \tag{4.177}$$

   where

$$E \exp(it\,\mathcal{F}) = e^{-t^2/2}\, E \exp(it\,\underline{\mathcal{F}}). \tag{4.178}$$

4. *Structure of the limit:*

$$\mathcal{F} = \mathcal{F}_0 + \underline{\mathcal{F}}, \tag{4.179}$$

   where for $(x,y) \in \mathbf{R}^2$

$$E \exp(ix\,\mathcal{F}_0 + iy\,\underline{\mathcal{F}}) = e^{-x^2/2}\, E \exp(iy\,\underline{\mathcal{F}}). \tag{4.180}$$

Proof. 1. This is a reformulation of weak convergence.

2. Using

$$E(Y_{n,1} \,|\, \underline{Y_n}) = \underline{Y_n}, \tag{4.181}$$

we evaluate the conditional variance in any row of the array,

$$E_n\{(Y_{n,1} - \underline{Y_n})^2 \,|\, \underline{Y_n}\} = 1 - \underline{Y_n}^2 \tag{4.182}$$

and obtain by virtue of eq.(4.175)

$$E_n\{(Y_{n,1} - \underline{Y_n})^2 \mid \underline{Y_n}\} \xrightarrow{\mathcal{D}} 1, \tag{4.183}$$

such that the limit is deterministic. Therefore

$$\begin{aligned}
&\lim_n E_n \exp\{it\frac{1}{\sqrt{n}} \sum_{i=1}^{n}(Y_{n,i} - \underline{Y_n})\} \\
&= \lim_n E_n \Big(1 - \frac{t^2}{2n} E_n\{(Y_{n,1} - \underline{Y_n})^2 | \underline{Y_n}\} + o(\frac{1}{n})\Big)^n \quad (4.184)\\
&= \exp(-\frac{t^2}{2}) = E \exp(it\,\mathcal{F}_0).
\end{aligned}$$

3. We evaluate the characteristic function

$$\begin{aligned}
&\lim_n E_n \exp\{it\,\frac{1}{\sqrt{n}} \sum_{i=0}^{n} Y_{n,i}\} \\
&= \lim_n E_n\Big(\exp(it\,\sqrt{n}\,\underline{Y_n})\; E_n\{\exp\Big(it\frac{1}{n}\sum^{n}(Y_{n,i} - \underline{Y_n})\Big) \mid \underline{Y_n}\}\Big) \\
&= \lim_n E_n\Big(\exp(it\,\sqrt{n}\,\underline{Y_n})\; \{1 - \frac{t^2}{2n} + o(\frac{1}{n})\}^n\Big) \quad (4.185)\\
&= \exp(-\frac{t^2}{2})\, E \exp(it\,\underline{\mathcal{F}}) = E \exp\{it\,(\mathcal{F}_0 + \underline{\mathcal{F}})\}
\end{aligned}$$

4. From part 2 and 3 we have

$$\begin{aligned}
E\{\exp(it\,\mathcal{F}_0) \mid \underline{\mathcal{F}}\} &= e^{-t^2/2}, \quad (4.186)\\
E\{\exp(it\,\mathcal{F}) \mid \underline{\mathcal{F}}\} &= e^{-t^2/2}\, E\, e^{it\,\underline{\mathcal{F}}},
\end{aligned}$$

such that $\mathcal{F} = \mathcal{F}_0 + \underline{\mathcal{F}}$ follows. By means of these conditional expectations we obtain

$$\begin{aligned}
&E \exp(ix\mathcal{F}_0 + iy\underline{\mathcal{F}}) \\
&= E\,(E\{\exp(ix\mathcal{F}_0 + iy\underline{\mathcal{F}}) \mid \underline{\mathcal{F}}\}) \\
&= E\,(\exp(iy\underline{\mathcal{F}})\, E\{\exp(ix\mathcal{F}_0) \mid \underline{\mathcal{F}}\}) \\
&= E \exp\{-x^2/2 + iy\,\underline{\mathcal{F}}\}. \quad (4.187)
\end{aligned}$$

QED

We start the discussion with remarks concerning part 2 and part 4 of the theorem.

Remark 4.3.11.

(a) Since the limit of the conditional variances is deterministic and normalized the limit (4.176) for the dependent centered random variables under consideration agrees with the usual limit for i.i.d. random variables. More general CLTs (with centering) for arrays of interchangeable random variables are considered in [40].

(b) $\mathcal{F}_0$ and $\mathcal{F}$ are statistically independent.

The following remarks refer to part 3 of the theorem.

Remark 4.3.12.

(a) The conditional probability of $\mathcal{F}$, given $\underline{\mathcal{F}}$, is determined by the formula

$$\frac{d}{dx} P(\mathcal{F} < x \,|\, \underline{\mathcal{F}}) = \frac{1}{\sqrt{2\pi}} \exp\{-\frac{(x - \underline{\mathcal{F}})^2}{2}\}. \tag{4.188}$$

Therefore the conditional expectation of $f(\mathcal{F})$ reads

$$E(f(\mathcal{F}) \,|\, \underline{\mathcal{F}}) = \int dx f(x) \frac{1}{\sqrt{2\pi}} \exp\{-\frac{(x - \underline{\mathcal{F}})^2}{2}\}. \tag{4.189}$$

(b) The structure of the mixtures determined by Theorem 4.3.6 entails the following expression for the variance

$$\mathrm{Var}(\mathcal{F}) = E\,\{\mathrm{Var}(\mathcal{F} \,|\, \underline{\mathcal{F}})\} + \mathrm{Var}\,\{E(\mathcal{F} \,|\, \underline{\mathcal{F}})\} \tag{4.190}$$

where the conditional variance is given by

$$\mathrm{Var}(\mathcal{F} \,|\, \underline{\mathcal{F}}) = E\,(\mathcal{F}^2 \,|\, \underline{\mathcal{F}}) - E^2\,(\mathcal{F} \,|\, \underline{\mathcal{F}}). \tag{4.191}$$

Using

$$E\,(\mathcal{F} \,|\, \underline{\mathcal{F}}) = \underline{\mathcal{F}}, \tag{4.192}$$

$$E\,(\mathcal{F}^2 \,|\, \underline{\mathcal{F}}) = 1 + \underline{\mathcal{F}}^2, \tag{4.193}$$

$$\mathrm{Var}(\mathcal{F} \,|\, \underline{\mathcal{F}}) = 1, \tag{4.194}$$

we obtain

$$\mathrm{Var}(\mathcal{F}) = 1 + \mathrm{Var}(\underline{\mathcal{F}}). \tag{4.195}$$

Therefore, if $E\mathcal{F} = E\underline{\mathcal{F}}$ is fixed, $\mathcal{F}$ has the smallest possible variance iff $\underline{\mathcal{F}}$ is deterministic.

(c) A direct proof of eq.(4.177) may be based upon the identity

$$E_n \exp(it\frac{1}{\sqrt{n}} \sum_{i=0}^{n} Y_{n,i}) = E_n\big(\cos(\frac{t}{\sqrt{n}}) + i\,\underline{Y_n} \sin(\frac{t}{\sqrt{n}})\big)^n. \tag{4.196}$$

The proof given above follows the strategy of ref. [1], namely on centering and compensation. This procedure shows in particular that the centered contributions and the sum of the mean values $\underline{Y_n}$ are conditionally independent in general and independent in the limit. Eqs.(4.175) and (4.176) imply

$$\begin{aligned} \frac{1}{\sqrt{n}} \sum^{n} Y_{n,i} &= \frac{1}{\sqrt{n}} \sum^{n} (Y_{n,i} - E_n Y_{n,i}) + \frac{1}{\sqrt{n}} \sum^{n} E_n Y_{n,i} \\ &= \frac{1}{\sqrt{n}} \sum_{i=1}^{n} (Y_{n,i} - E_n Y_{n,i}) + \sqrt{n}\,\underline{Y_n} \overset{\mathcal{D}}{\to} \mathcal{F}_0 + \underline{\mathcal{F}}, \end{aligned} \tag{4.197}$$

but do not show that $\mathcal{F}_0$ and $\underline{\mathcal{F}}$ are independent.

(d) Since the limit of the conditional variances is deterministic the mixing of the Gaussian distributions under consideration concerns only the mean values and not the variance.

(e) There is a connection between the the strategy of not centering and the deterministic conditional variance that is given by the implication

$$\sqrt{n}\,\underline{Y_n} \overset{\mathcal{D}}{\to} \underline{\mathcal{F}} \quad \Rightarrow \quad E_n\Big((Y_{n,1} - \underline{Y_n})^2 | \underline{Y_n}\Big) \overset{\mathcal{D}}{\to} 1. \tag{4.198}$$

Lemma 4.3.8. *For any $n \in \mathbf{Z}_+$ (i)*

$$E(\mathcal{F}^n | \underline{\mathcal{F}}) = \tilde{H}_n(\underline{\mathcal{F}}), \tag{4.199}$$

where

$$\tilde{H}_n(x) = \sum_{k=0}^{[n/2]} \frac{n!}{k!(n-2k)!} \left(\frac{1}{2}\right)^k x^{n-2k}, \tag{4.200}$$

and (ii)

$$E\Big(H_n(\mathcal{F})\,|\,\underline{\mathcal{F}}\Big) = \underline{\mathcal{F}}^n, \tag{4.201}$$

where

$$H_n(x) = \sum_{k=0}^{[n/2]} \frac{n!}{k!(n-2k)!} \left(-\frac{1}{2}\right)^k x^{n-2k} \tag{4.202}$$

denote the Hermite polynomials.

Proof. For any $n \in \mathbf{Z}_+$ the expansion of $\mathcal{F}^n$, given $\underline{\mathcal{F}}$, is determined by

$$E(\mathcal{F}^n | \underline{\mathcal{F}}) = \sum_{k=0}^{n} \binom{n}{k} E(\mathcal{F}_0^{n-k})\, \underline{\mathcal{F}}^k. \tag{4.203}$$

Using, for $n \in \mathbf{Z}_+$, the formulae

$$E(\mathcal{F}_0^{2n}) = \frac{(2n)!}{n!2^n}, \qquad E(\mathcal{F}_0^{2n+1}) = 0 \tag{4.204}$$

we obtain (i). Comparing this with the inversion formula

$$x^n = \sum_{k=0}^{[n/2]} \frac{n!}{k!(n-2k)!} \left(\frac{1}{2}\right)^k H_{n-2k}(x) \tag{4.205}$$

where we replace x by $\mathcal{F}$ and evaluate the conditional expectation gives (ii).

QED

Remark 4.3.13. Using the results of [21] we obtain the following quantum identities related to the transformation to normal order (cf. e.g. [80, Prop. 1.5.5] for part (ii)).

On the domain of finite linear combinations of the eigenvectors $\Phi(k)$ of the number operator a^+a or evaluated by the linear functionals $E_{z,z'}, z, z' \in \mathbf{C}$

$$E_{z,z'}(\cdot) = < \Psi(z), \cdot > \Psi(z'), \tag{4.206}$$

where $\Psi(z), z \in \mathbf{C}$, is a coherent state vector, for all $n \in \mathbf{N}$ and all $\alpha \in \mathbf{C}, |\alpha| = 1$,

(i)

$$(\alpha a^+ + \overline{\alpha} a)^n = \sum_{k=0}^{[n/2]} \frac{n!}{k!(n-2k)!} (\frac{1}{2})^k \sum_{j=0}^{n-2k} \binom{n-2k}{j} (\alpha a^+)^j (\overline{\alpha} a)^{n-2k-j}, \tag{4.207}$$

and (ii)

$$H_n\left(\alpha a^+ + \overline{\alpha} a\right) = \sum_{k=0}^{n} \binom{n}{k} (\alpha a^+)^k (\overline{\alpha} a)^{n-k}. \tag{4.208}$$

For the quantum generalization of the central limit of de Finetti's theorem we refer to [21]. In the physical interpretation given there $\mathcal{F}$ denotes a (magnetic or electric) field observable, $\mathcal{F}_0$ the vacuum field and $\underline{\mathcal{F}}$ the excitation field such that the field is the sum of the vacuum field and the excitation field. The conditional expectation

$$E\{ f(\mathcal{F}) \,|\, \underline{\mathcal{F}} \} = E\{ f(\mathcal{F}_0 + \underline{\mathcal{F}}) \,|\, \underline{\mathcal{F}} \} \tag{4.209}$$

eliminates the vacuum field $\mathcal{F}_0$, which historically is the origin of the transformation to normal order in quantum electrodynamics.

We conclude this section with two problems.

Problem 4.3.2. What are the sufficient and necessary extendibility conditions for the convergence to a mixture of Gaussian distributions if the array of interchangeable $\{-1, 1\}$–valued random variables is not ∞–extendible?

Problem 4.3.3. Are there generalizations of the Poisson limit and the central limit for $\delta \notin \{0, 1/2\}$?

4.3.3 Large Deviations

Besides the LLN, the CLT and the Poisson limit theorem, large deviation (LD) properties play a fundamental role in probability theory (see e.g. [63, 40]) and statistical mechanics [60]. The entropies that are determined by large deviations establish a canonical link between probability and statistical mechanics and have been analyzed, in general, for i.i.d. random variables. Although some generalizations for interchangeable random variables exist [25, 26, 52], the applications to the statistics of indistinguishable particles [145, 146, 147] are not conclusive.

5. Historical and Conceptual Remarks

In this concluding chapter specific historical and conceptual topics pertaining to our theme are considered. These sections are no substitute for a historical or conceptual analysis. Rather, I consider specific situations connected with the misevolution of the concept of identical and indistinguishable objects. Moreover, the origin of the terminological misevolution is explicitly identified. Finally, I consider the notion of 'negative probability' as it is used in quantum optics.

In general, I refrain from any discussion of the existing historical and conceptual literature because of the special viewpoint (permutation invariance of the state) considered here. As far as the contributions of BOLTZMANN and of EINSTEIN are concerned, I refer the reader to my articles 'Boltzmann's probability distribution of 1877' [15] and 'Eine Fehlinterpretation mit Folgen: Albert Einstein und der Welle–Teilchen Dualismus' [12].

5.1 Historical Remarks

5.1.1 Identity, Indistinguishability and Probability

The existence of identical particles is the fundamental assumption of all variants of atomism from antiquity until today. Prior to the description of a state by means of probability measures states were identified with point measures. In this deterministic setting indistinguishable objects are not conceivable (cf. Section 1.2). One general statement concerning this fact is LEIBNIZ's *principium identitas indiscernibilium* (see e.g. [94, 153])

This principle expresses that 'there is no such thing as two individuals indiscernible from each other' or 'non dari posse in natura duas res singulares solo numero differentes' (see [94]). In modern terms: 'For, to count them [particles] as two or more there should be something that distinguishes them' [115, p. 121]. This principle refers to a deterministic setting and denies the existence of identical – and therefore also of indistinguishable – objects if both intrinsic and – in particular – extrinsic properties are considered.

This viewpoint changed partly under the influence of the natural philosophers of the following centuries. For the atomists the existence of identical objects is a necessary assumption whereas indistinguishable objects are asserted not to exist. As explained by LANGE: 'Du Bois's famous 'iron particle', which definitely is the same 'thing', whether it passes in a meteor stone through the universe, rushes in a steam locomotive's wheel on the rails, or runs in a blood cell through the poet's temple, is in all these cases just precisely the 'same thing' because we refrain from considering the peculiarity of its position with respect to other particles and the interactions following therefrom ...' [107, p. 659]. And identity is the basis of atomism today: 'First, as far as we know, any two particles of the same species are, except for their position and state of motion, absolutely identical, whether they occupy the same atom or lie at the opposite ends of the Universe.' [152, p. 50]

Indistinguishability, by contrast, is a rather recent concept. It is a notion which unnoticeably is introduced into the physical formalism by means of a probabilistic description of states by BOLTZMANN. In this context it is essential that prior to DE FINETTI's investigations the concept of independent and identically distributed random variables has not been clearly identified with invariance under permutations. At least from the physical viewpoint, the fact that independent identical repetitions of an experiment are indistinguishable repetitions has not been recognized. The irrelevance of order, underlying indistinguishability, seems so self–evident that it remained unnoticed.

In this context I would like to recall that all conceptual difficulties concerning the notion of indistinguishability are present in the familiar description of tossing two identical coins by two i.i.d. random variables. Mathematical textbooks are usually not very explicit to explain why the sum of two copies of a random variable $X : \Omega \to \{H,T\}$ is not equal to $2X$ but equal to $X_1 + X_2$. Precisely by this construction (a copy is another random variable with the same distribution), however, classical probability establishes a description of *different* objects that are *indistinguishable*.

5.1.2 Boltzmann and Bose–Einstein Statistics

To demonstrate the compatibility of the assumption of a microscopic atomistic structure with macroscopic continuous phenomena, BOLTZMANN derives in 1868 [33] the Maxwell velocity distribution in two dimensions transformed to the energy $\mathcal{E}$ (see eq.(3.154)), $\kappa = E\,\mathcal{E} > 0$,

$$\frac{d}{dx} P(\mathcal{E} < x) = \frac{1}{\kappa} \exp(-\frac{x}{\kappa}), \quad x \geq 0, \tag{5.1}$$

from a discrete setting. To this end he introduces the scheme where n particles (energy elements ε) are distributed onto d cells (molecules). Assuming all distributions of the n identical energy elements onto the d molecules to be *a priori* equal probable, he evaluates the cardinality of the set of sequences of occupation numbers and obtains the probability distribution (see eq.(3.68))

$$P_{BE}(\mathbf{K} = \mathbf{k}) = \binom{d+n-1}{n}^{-1}. \tag{5.2}$$

Second, he determines the marginal distribution of the number of energy elements ε in an arbitrarily chosen cell (see eq.(3.125))

$$P_{BE}(K=k) = \binom{d+n-1}{n}^{-1} \binom{d+n-k-2}{n-k}. \tag{5.3}$$

Third, BOLTZMANN takes the continuum limit for the quantity, $d\kappa = n\varepsilon$,

$$\mathcal{E}_{n,d} = d\,\kappa\, K/n \tag{5.4}$$

and obtains by means of the limit $n \to \infty$ (this is the LLN) a scaled beta-distribution (see eq.(4.25))

$$\frac{d}{dx} \lim_{n\to\infty} P_{BE}(\mathcal{E}_{n,d} < x) = p_d(x) = \frac{d-1}{d\kappa}(1 - \frac{x}{d\kappa})^{d-2}. \tag{5.5}$$

Eventually, in the macroscopic limit $d \to \infty$ he obtains the exponential distribution (5.1).

Several years later, in a reply to Loschmidt's 'reversibility paradox', BOLTZMANN argues that the macroscopic thermostatic properties of a system can be deduced from the most probable state of a microscopic version of the system, an idea which is the subject of his famous memoir of 1877 [34]. In this memoir discrete symmetric probabilities became the foundation of the theory. BOLTZMANN distributes n energy elements onto d molecules, introduces occupation numbers (complexions) and occupancy numbers. The fundamental probability distribution is is the uniform distribution of the occupation numbers (BE statistics). Based upon this assumption BOLTZMANN evaluates the distribution of the occupancy numbers (see eq.(3.71))

$$P_{BE}(\mathbf{Z}=\mathbf{z}) = \binom{d}{z_0 \ldots z_n} \binom{d+n-1}{n}^{-1}. \tag{5.6}$$

To determine the most probable state $\mathbf{z}^*$ BOLTZMANN minimizes the expression

$$M(\mathbf{z}) = \sum_{i=0}^{n} \ln(z_i!) \approx \sum_{i=0}^{n} z_i \ln z_i \tag{5.7}$$

subject to the constraints

$$\sum_{i=0}^{n} z_i = d, \quad \sum_{i=0}^{n} i\, z_i = n. \tag{5.8}$$

From this assumption he derives from BE statistics in the continuum limit the fundamental result that the limit of the combinatorial entropy $M(\mathbf{z}^*)$ is equal (up to an additive constant and a factor) to the entropy of phenomenological thermostatics in two dimensions.

On the other hand, he shows that in three dimensions the same result follows if one uses instead of the occupancy numbers in BE statistics the occupation numbers in MB statistics. It seems, however, that the shift in the scheme of levels and the exchange of probability distributions, was not noticed by BOLTZMANN himself because several pages later, in the context of a critique, he asserts that besides his choice of the fundamental probability distribution, eq.(5.6), another probability distribution is quite natural and conceivable. By means of the urn model with replacement he introduces MB statistics explicitly. Since the Poisson limit theorem is unknown, he obtains no macroscopic and continuum limit. Consequently, he determines by combinatorial tools the distribution of the occupancy numbers, obtains eq.(3.70)

$$P_{MB}(\mathbf{Z} = \mathbf{z}) \;=\; \begin{pmatrix} & d & \\ z_0 & \ldots & z_n \end{pmatrix} \frac{n!}{0!^{z_0} \cdots n!^{z_n}} \left(\frac{1}{d}\right)^n \tag{5.9}$$

and remarks that the limit of this probability distribution (MB statistics) yields not the thermal equilibrium state.

5.1.3 Einstein and the Wave–Particle Duality

Let us recall, before going into the historical details, the macroscopic limit of the particle number $\mathcal{M}$ of a group containing d cells in BE statistics which is given by a negative binomial distribution $n_{\overline{n},d}$ with integral representation

$$P(\mathcal{M} = k) \;=\; \int dx \, (\overset{d}{\star}\, \Gamma_{\frac{1}{\overline{n}},1})(x)\, \pi_x(k). \tag{5.10}$$

Defining the energy of the system under consideration by $\mathcal{E} = \varepsilon \mathcal{M}, \varepsilon > 0$, this yields the variance formula

$$\mathrm{Var}(\mathcal{E}) \;=\; \varepsilon\, E(\mathcal{E}) + \frac{1}{d}\, E^2(\mathcal{E}). \tag{5.11}$$

Formula (5.11) has two distinguished limits.

1. For $d \to \infty, \overline{n} \downarrow 0$ such that $d\overline{n} \to \delta > 0$ we obtain (this defines $\mathcal{M}_p$)

$$P(\mathcal{M} = k) \to P(\mathcal{M}_p = k) \;=\; \pi_\delta(k) \tag{5.12}$$

with the corresponding variance formula for $\mathcal{E}_p = \varepsilon \mathcal{M}_p$

$$\mathrm{Var}(\mathcal{E}_p) \;=\; \varepsilon\, E(\mathcal{E}_p). \tag{5.13}$$

2. On the other hand, for $\varepsilon \downarrow 0, \overline{n} \to \infty$ such that $\varepsilon \overline{n} \to \kappa > 0$ we obtain (this defines $\mathcal{E}_w$)

$$P(\mathcal{E} < x) \to P(\mathcal{E}_w < x) \;=\; \int_0^x dy \Gamma_{\frac{1}{\kappa},d}(y) \tag{5.14}$$

with the variance formula

$$\mathrm{Var}(\mathcal{E}_w) \;=\; \frac{1}{d}\, E^2(\mathcal{E}_w). \tag{5.15}$$

In each of these limits precisely one term of the variance formula (5.11) survives as a function of the mean value. This, however, does not imply that – for independent $\mathcal{E}_p$ and $\mathcal{E}_w$ – the relation $\mathcal{E} = \mathcal{E}_p + \mathcal{E}_w$ holds (cf. the calculations in [12]).

Analyzing the entropy density of black-body radiation in the limit of large frequencies, EINSTEIN [55] realizes in 1905 a structural analogy between this system and the ideal gas. To explain this analogy he introduces the light quantum hypothesis, namely, that black-body radiation (in this limit) consists of a system of statistically independent energy elements obeying the energy–frequency relation $\varepsilon = \hbar\omega$. In 1909 he generalizes this analysis [56], starting from Planck's law

$$E(\mathcal{E}) = d\,\hbar\omega \frac{1}{e^{\beta\hbar\omega} - 1}. \tag{5.16}$$

Here $\mathcal{E}$ denotes the energy density (with respect to ω) and $d = d(\omega)$ denotes the density of modes of the electromagnetic field. In particular, he determines the variance of $\mathcal{E}$ (cf. eq.(3.139))

$$\mathrm{Var}(\mathcal{E}) = \hbar\omega\, E(\mathcal{E}) + \frac{1}{d} E^2(\mathcal{E}) \tag{5.17}$$

as a function of the mean value. For the interpretation EINSTEIN considers both terms *separately*. The first term, he argues, can be explained by the light quantum hypothesis (assuming the quanta to be distributed according to a Poisson distribution). The second term, he argues, can be explained from the wave properties of the electromagnetic field. This argument was confirmed by LORENTZ [111] in 1916 who shows, using the CLT, that the energy density is distributed according to a $\Gamma_{1/\overline{n},d/\overline{n}}$ distribution with mean $\overline{n}$. For the distribution of both terms in eq.(5.17) *together* EINSTEIN refers to Bienaymé's identity, that is, he assumes that the energy density $\mathcal{E}$ of blackbody radiation consists of a 'particle' part $\mathcal{E}_p$ and a 'wave' part $\mathcal{E}_w$

$$\mathcal{E} = \mathcal{E}_p + \mathcal{E}_w, \tag{5.18}$$

and that these constituents are statistically independent and contribute to the variance by the two terms given in eq.(5.17). By this interpretation he establishes the wave-particle dualism of light.

In 1915 VON LAUE [148] is able to show that the random variable underlying Planck's law and Einstein's fluctuation equation is the sum of independent identically distributed copies of a random variable $\mathcal{N}$ with a geometric distribution

$$\mathcal{E} = \hbar\omega \sum_{i=1}^{d} \mathcal{N}_i. \tag{5.19}$$

This allows him to deduce the variance (5.17) of $\mathcal{E}$ from the variance of $\mathcal{N}$

$$\mathrm{Var}(\mathcal{N}) = E(\mathcal{N}) + E^2(\mathcal{N}). \tag{5.20}$$

In 1924-25 EINSTEIN applies BOSE's statistics to the ideal quantum gas and obtains for the variance of the energy of the system again the two terms of eq.(5.20) (see eq.(3.139 and eq.(4.120)). Whereas the first term, the 'particle' term, was expected to exist for particles, the second term had to be explained. Generalizing his dualistic light theory to particles, EINSTEIN [59] associates, to account for the 'mystical interdependence' of the particles, a wave with the particles, thus establishing an universal wave–particle dualism.

EINSTEIN's wave–particle duality is an interpretation of the variance of the geometric distribution which, however, is unacceptable for the following reasons.

- Because $\mathcal{N}$ is a non-negative integer-valued random variable, a dualistic interpretation of the variance is impossible. Partly this conclusion is drawn by BOTHE [36] who establishes in 1927 the de Finetti-type integral representation (see eq.(4.115)), $\overline{n} = E\mathcal{N}$,

$$P(\mathcal{N} = k) = \int dx \, \frac{1}{\overline{n}} \exp(-x/\overline{n}) \, \pi_x(k) \tag{5.21}$$

  and, based upon this representation, concludes that the two terms of eq.(5.20) are interdependent.
- According to remark **(f)** p. 115 the first term in eq.(5.20) is a general contribution whereas the second term is a specific contribution generated precisely by a statistical dependence among the objects. In particular, the two terms cannot be interpreted as due to two independent additive sources.

EINSTEIN's wave-particle duality is the interpretation of a mathematical formula, however, – as shown here – it is a misinterpretation. BOHR, by contrast, never formalized his duality concept and was, therefore, able to generalize it to complementarity. In view of EINSTEIN's misinterpretation the wave nature of particles (in the framework of early quantum statistics) is merely a metaphor for statistical dependence of particles.

5.1.4 Quantum Theory

PAULI's exclusion principle states that the sets of four quantum numbers of any two electrons of an atom must be different. In this naive form the exclusion principle suggests that electrons are distinguishable, namely identifiable by their quantum numbers. The new, probabilistic, aspect is formulated by FERMI [64] and implies negative correlations. The particles of BE statistics, by contrast, have – up to exceptions (antibunching) – positive correlations which are well known from the photon bunching of thermal light.

The transition to quantum theory starts with HEISENBERG's [88] and DIRAC's [53] quantum investigations and reveals the fundamental meaning of permutation invariance of the state in terms of symmetric and antisymmetric wave functions. The application of this symmetry soon gives an explanation of the constitution of atoms, of the properties of their spectra and of the nature of the chemical bond. The latter purely statistical phenomenon is interpreted, however, by HEITLER and LONDON [89] in terms of an 'exchange–interaction' thus intermingling the clearly separated domains of states and dynamics.

After the invention of the formalism of quantum theory the theory of indistinguishable particles splits into two incompatible domains, a classical part where indistinguishability is defined by the combinatorial indistinguishability concept of early quantum theory (see Section 5.2 below) and a quantum part where indistinguishability is defined by means of anti/symmetry of wave functions. Meanwhile, under the influence of WIGNER [154] and WEYL [153] the emphasis switches from the anti/symmetry of wave functions to the symmetry of observables and the representation theory of the symmetric group for these. Moreover, the familiar definition of indistinguishability by means of the symmetry of the observables is more or less explicitly based on the assumption that, according to the Copenhagen interpretation, quantum theory is a formalism for the description of *observations* of a physical system.

Finally, since 1930 there exists a third independent mathematical branch, founded – in the framework of subjective probability – by DE FINETTI. From the quantum viewpoint this branch analyzes symmetric states in abelian subalgebras. This line of investigations starts with the strong LLN, establishes a $0-1$ law [90] and derives the equivalence of ∞–extendibility and conditional identity and independence [109]. It eventually unifies the quantum and the classical domain in the framework of C^*–algebras by the quantum de Finetti theorem [139].

One of the best investigated phenomena, showing the consequences of indistinguishability, are scattering processes of indistinguishable particles, first analyzed by MOTT [122]. In this context, FEYNMAN [65] proposes an interpretation of the superposition principle which is related but not identical with the notion of indistinguishability. According to FEYNMAN amplitudes (wave functions) for indistinguishable alternatives add. Since here indistinguishability is not defined by permutation invariance and since this prescription also applies to one particle, this concept of indistinguishability refers to situations where fundamental events cannot be distinguished. As an example I recall the two–slit experiment or interference phenomena of dynamically (not necessarily statistically) independent light beams [126]. For indistinguishable alternatives (paths) quantum theory postulates *a priori* correlations.

5.2 The Combinatorial Concept of Indistinguishability

The concept of indistinguishable particles was analyzed for the first time in 1911 by NATANSON [124] as a methodological tool to understand the difference between MB and BE statistics (the latter is BOLTZMANN's BE statistics of 1877). In this context it is important to emphasize that BOLTZMANN introduced MB statistics at the three levels (configurations, occupation numbers, occupancy numbers) whereas for BE statistics he did not mention the configurations.

To explain NATANSON's analysis let us assume that n identical particles are distributed onto d cells and that an occupancy number $\mathbf{z}$ is given. Then, according to NATANSON, neither the cells nor the particles are identifiable, since no information exists which would allow us to infer which particle is in which cell, nor would we be able to ascertain which one of the cells contains just $k_i, 1 \leq i \leq d$, particles. On the other hand, if an occupation number $\mathbf{k}$ is given, then the cells are identifiable, since the cells may be identified by the number of particles they contain (this presupposes that all occupation numbers are different). The particles, however, are still indistinguishable. According to NATANSON the particles are distinguishable if a configuration $\mathbf{j}$ is given.

Whereas this analysis refers to events, NATANSON applies this argument not to events but to probabilities of events assumed to be uniformly distributed and defines indistinguishable and distinguishable particles as follows.

Definition 5.2.1. *Identical particles of the statistical scheme are called* combinatorially distinguishable *if configurations are defined and uniformly distributed. They are called* combinatorially indistinguishable *if configurations are not defined and the occupation numbers are uniformly distributed.*

Remark 5.2.1.

(a) According to this definition the particles of MB statistics are combinatorially distinguishable and the particles of BE statistics (as used by BOLTZMANN 1877, PLANCK 1901 [130] and LORENTZ 1910 [110]) are combinatorially indistinguishable. Since all derivations of Planck's law are based upon the assumption of equal *a priori* probabilities for the occupation numbers (BE statistics), NATANSON concludes that the combinatorial indistinguishability of the particles is the foundation of PLANCK's theory.

(b) To incorporate FD statistics into this scheme it is necessary to assume that events with more than one particle in a cell do not exist.

(c) According to NATANSON, the uniform distribution of the quantities (configurations, occupation numbers, occupancy numbers) on one level of the statistical scheme guarantees their identifiability and entails the indistinguishability of the quantities on the levels below it. By our definition, however, uniformly distributed identical objects are indistinguishable themselves, and this implies the indistinguishability of the quantities on the lower levels. Moreover, according to NATANSON, indistinguishability is a consequence of equal *a priori* probabilities. By our definition, however, indistinguishability is equivalent to the invariance of probabilities under permutations, and the uniform distribution is only a trivial example of a symmetric probability measure.

(d) Conceptually, NATANSON's investigation is confined to uniform distributions and depends, therefore, on a combinatorial analysis of the cardinality of the fundamental events. The determination of this number by means of the familiar picture with particles in boxes has been confused in general with the goal of this computation, namely, the determination of a probability distribution (that is invariant under permutations).

NATANSON's (combinatorial) concept of indistinguishability was immediately accepted by PLANCK [131], and since then it has survived in familiar physical textbooks as the conceptual foundation for a misleading terminology. Since then, this definition of indistinguishability is used in the textbooks on classical probability theory (see e.g. [62, 97]) and combinatorics (see e.g. [128]). NATANSON's analysis and his terminology are quite natural prior to the establishment of quantum theory. After the discovery, in this context, of the fundamental meaning of permutation invariance, this terminology is unfounded.

In this context it may be useful to recall a remark concerning the application of the combinatorial concept of indistinguishability in probability theory. FELLER remarks: '*Whether or not actual balls are in practice distinguishable is irrelevant for our theory.* Even if they are, we may decide to treat them as indistinguishable' [62, p. 12].

5.3 'Negative Probabilities' in Quantum Optics

By definition, probabilities are non-negative quantities. Nonetheless, physicists have introduced, in particular for the 'explanation' of certain quantum phenomena (EPR paradox, violations of Bell–type inequalities), the concept of 'negative probabilities' (for a review see [123]). In my opinion the introduction of 'negative probabilities' is self–contradictory and methodologically misleading insofar as it suggests that a precise analysis by means of established concepts is impossible.

5.3.1 Signed Binomial and Poisson Mixtures

For unextendible sequences of $\{0,1\}$–valued interchangeable random variables de Finetti's classical theorem does not hold so that an integral representation in terms of the binomial distribution does not exist. There exists, however, a representation in terms of hypergeometric distributions (see Sections 3.2 and 4.1). Here we show that any probability distribution on $\{0,1,\ldots,n\}$ admits a representation in terms of the binomial distribution by means of a (not uniquely determined) signed measure on $[0,1]$.

Lemma 5.3.1. *For any probability distribution* $\mathbf{p} = (p_0,\ldots, p_n)$ *on* $\{0,1,\ldots,n\}$ *there exists a signed measure* $\sigma \in M_b^1([0,1])$, *the set of bounded measures on* $[0,1]$, *normalized according to* $\int d\sigma = 1$, *such that for any* $k, 0 \leq k \leq n$, *the representation*

$$p_k = \int d\sigma(q)\, B_{n,q}(k) \tag{5.22}$$

holds.

Proof. For any $n \in \mathbf{N}$ the Bernstein polynomials $B_{n,q}(k) \in C([0,1]), 0 \leq k \leq n$, are linear independent such that for any choice of $\mathbf{q} = \{q_0, \ldots, q_n\}$ where $0 \leq q_0 < q_1 < \cdots < q_n \leq 1$ the $n \times n$ matrix

$$\beta_{i,k} = B_{n,q_i}(k) \tag{5.23}$$

is invertible. Accordingly, the linear equation

$$p_k = \sum_{i=0}^{n} \sigma_i\, B_{n,q_i}(k) \tag{5.24}$$

always has a nontrivial solution $\underline{\sigma} \in \mathbf{R}^n$. Obviously,

$$1 = \sum_{k=0}^{n} p_k = \sum_{i=0}^{n} \sigma_i \sum_{k=0}^{n} B_{n,q_i}(k) = \sum_{i=0}^{n} \sigma_i, \tag{5.25}$$

such that $\underline{\sigma}$ is a normalized discrete signed measure. In particular, this shows that the signed measures are not uniquely determined. QED

Remark 5.3.1.
(a) JAYNES [96] gives a different proof, constructing for any $\mathbf{p}$ a polynomial of order n which acts as the density of a signed measure.
(b) The generalization to the multivariate case (representations in terms of the multinomial distribution by means of signed measures) is obvious and omitted here.
(c) These representations are connected with quantum representations of statistical operators on $\overset{n}{\otimes}\, \mathbf{C}^2$ by means of a signed measure in terms of discrete coherent states (cf. Section 2.3).

Since the functions $\pi_x(k) \in C_b(\mathbf{R}_+), k \in \mathbf{Z}_+$, are linear independent, by means of the same strategy one proves the following generalization.

Lemma 5.3.2. *For any probability distribution $\Pi \in M_+^1(\mathbf{Z}_+)$ there exists a normalized signed measure $\sigma \in M_b^1(\mathbf{R}_+)$, such that for any $k \in \mathbf{Z}_+$ the representation*

$$\Pi(k) = \int d\sigma(x)\, \pi_x(k) \tag{5.26}$$

holds.

5.3.2 The P–Representation

In quantum optics 'negative probabilities' have been introduced in the context of the P–representation of statistical operators of the quantum harmonic oscillator (see [79]). It is the goal of this final section to analyse this phenomenon and to elucidate that these 'negative probabilities' have no physical meaning. The Hilbert space $\mathcal{H}$ under consideration is the closed linear span of the c.o.n.s. of number vectors $\Phi(k), k \in \mathbf{Z}_+$. By $\Psi(z), z \in \mathbf{C}$, we denote the coherent state vectors (see eq.(2.71)).

Theorem 5.3.1. *For any $W \in \mathcal{S}(\mathcal{H})$ there exists a sequence $\sigma_n \in M_b(\mathbf{C})$ such that for any $B \in \mathcal{B}(\mathcal{H})$ we have*

$$tr(W\,B) = \lim_n \int d\sigma_n(z) \; < \Psi(z),\, B\,\Psi(z) > . \tag{5.27}$$

Proof. According to [43] the set of finite real linear combinations of coherent states

$$\sum_{i=1}^{n} \lambda_i < \Psi(z_i),\, \cdot > \Psi(z_i) \tag{5.28}$$

$\lambda_i \in \mathbf{R}, 1 \leq i \leq n$, is dense in the Banach space of self–adjoint trace class operators in the trace norm topology. Accordingly for any $W \in \mathcal{S}(\mathcal{H})$ there exists a sequence $W_n \in \mathcal{S}(\mathcal{H})$

$$W_n = \sum_{i=1}^{n} \lambda_{n,i} < \Psi(z_{n,i}),\, \cdot > \Psi(z_{n,i}) \tag{5.29}$$

such that for any $B \in \mathcal{B}(\mathcal{H})$ we have

$$\lim_n tr(\,W_n\,B\,) = tr(\,W\,B\,). \tag{5.30}$$

Setting

$$W_n = \int d\sigma_n(z) < \Psi(z),\, \cdot > \Psi(z) \tag{5.31}$$

where

$$\sigma_n = \sum_{i=1}^{n} \lambda_{n,i}\, \delta_{z_{n,i}} \tag{5.32}$$

determines the representation (5.28). QED

There exist two strategies for the representation of an arbitrary statistical operator in terms combinations of coherent states.

1. For any $B \in \mathcal{B}(\mathcal{H})$ the function $f_B : \mathbf{C} \to \mathbf{C}$ defined by
$$f_B(z) = <\Psi(z), B\Psi(z)> \tag{5.33}$$
is entire analytic and vanishing at infinity. The dual space of these functions is a space of distributions such that one obtains a representation of states in terms of not necessarily positive distributions which may be more singular than a Dirac measure (cf. e.g. [102, 120]).
2. Since $f_B(z) \in C_b(\mathbf{C})$ one can try to represent statistical operators W by a normalized signed measure $\sigma_W \in M_b^1(\mathbf{C})$.

Problem 5.3.1. Exists for any $W \in \mathcal{S}(\mathcal{H})$ a signed measure $\sigma_W \in M_b^1(\mathbf{C})$ such that for any $B \in \mathcal{B}(\mathcal{H})$ the representation
$$\mathrm{tr}(W\,B) = \int d\sigma_W(z)\ <\Psi(z), B\Psi(z)> \tag{5.34}$$
holds?

Whenever a representation in terms of a signed measure σ_W exists we recover eq.(5.26) in the abelian subalgebra generated by the number operator
$$\Pi(k) = \int d\sigma_W(z)\,\pi_{|z|^2}(k). \tag{5.35}$$

Moreover, in this case there exists a representation of probability densities on $\mathbf{R}$ in terms of linear combinations of Gaussian distributions (in the abelian subalgebra generated by the coordinate Q or the momentum P)
$$p_Q(x) = \int d\sigma_W(z)\,\frac{1}{\sqrt{2\pi}}\exp\{-\frac{(x-2\mathrm{Re}(z))^2}{2}\}, \tag{5.36}$$
$$p_P(x) = \int d\sigma_W(z)\,\frac{1}{\sqrt{2\pi}}\exp\{-\frac{(x-2\mathrm{Im}(z))^2}{2}\}. \tag{5.37}$$

The existence of these representations by means of objects which are no probability measures (signed measures or distributions which are more singular than a Dirac measure) is usually interpreted as a typical quantum phenomenon, transcending classical conceptions. According to the spectral theorem, however, for any statistical operator W of the harmonic oscillator there exists an uniquely determined probability distribution $\Pi \in M_+^1(\mathbf{Z}_+)$ such that
$$<\Phi(k), W\Phi(k)> = \Pi(k) \tag{5.38}$$
holds for all $k \in \mathbf{Z}_+$. In particular, whenever an integral representation for the state exists, the spectral theorem transfers this representation to any abelian subalgebra. Accordingly, whenever the representing distribution σ_W in eq.(5.28) is no probability measure, then this representation is artificial and the representing distribution is without any probabilistic significance. Whereas the de Finetti measures of Theorems 4.3.2, 4.3.3 have a precise

physical meaning (according to the renormalized LLN), the signed measures of Lemma 5.3.2 do not correspond to the distribution of any physical observable.

To summarize, the P–representation of statistical operators (photon states) can be divided into two classes (cf. [20]).

- Whenever the representing object is a probability measure, this probability measure has a well defined physical meaning by virtue of the quantum Poisson and central limit of de Finetti's theorem [21]. Classical states are the limit of a sufficiently fast extendible or an ∞–extendible array of quanta.
- If the representing object is not a probability measure the representation is artificial. The existence of nonclassical states is not a specific quantum phenomenon transcending classical behaviour. Rather, the nonexistence of an integral representation is due to the unextendibility of the underlying array of quanta. Nonclassical photon states are the limit states of a not sufficiently fast extendible array of quanta and the specific correlation properties of nonclassical light are due to this insufficient speed of extendibilty (see the examples for number states p. 122 and squeezed states p. 122 in Section 4.3).

A. Probability Distributions and Notation

A.1 Probability Distributions

The quantities in the first line, following the name of the probability distribution, are the independent variables. The quantities in the next lines are the parameters. The notation agrees as far as possible and convenient with that by FELLER [62, 63].

Binomial distribution

$k \in \mathbf{Z}_+, 0 \le k \le n,$

$p \in [0,1],$

$$B_{n,p}(k) = \binom{n}{k} p^k (1-p)^{n-k}. \tag{A.1}$$

Hypergeometric distribution

$k \in \mathbf{Z}_+, 0 \le k \le \min(r,m),$

$m,n,r \in \mathbf{Z}_+, m,r \le n,$

$$\mathcal{H}_{n,m,r}(k) = \binom{n}{r}^{-1} \binom{m}{k} \binom{n-m}{r-k} = \mathcal{H}_{n,r,m}(k). \tag{A.2}$$

Multinomial distribution

$\mathbf{k} \in \{0,1,\ldots,n\}^d, \sum_{i=1}^d k_i = n,$

$0 \le p_i \le 1, 1 \le i \le d, \sum_{i=1}^d p_i = 1,$

$$M_{n,\mathbf{p}}(\mathbf{k}) = \binom{n}{k_1 \ldots k_n} p_1^{k_1} \cdots p_n^{k_n}. \tag{A.3}$$

Polyhypergeometric distribution

$\mathbf{n} \in \{0,1,\ldots,n\}^f, \sum_{i=1}^f n_i = n,$

$\mathbf{r} \in \{0,1,\ldots,m\}^f, \sum_{i=1}^f r_i = m,$

$m \le n, n_i \le r_i, i \le i \le f,$

$$\mathcal{PH}_{n,m,\mathbf{r}}(\mathbf{n}) = \binom{n}{m}^{-1} \prod_{i=1}^{f} \binom{r_i}{n_i}. \tag{A.4}$$

Pólya-Brillouin distribution

$\mathbf{n} \in \{0, 1, \ldots, n\}^f, \sum_{i=1}^f n_i = n,$
$\mathbf{d} \in \{1, 2, \ldots, d\}^f, \sum_{i=1}^f d_i = d,$
$c \in \mathbf{R},$

$$P(\mathbf{N} = \mathbf{n}) = \left(\begin{array}{c} n \\ n_1 \ \ldots \ n_f \end{array} \right) \{(\frac{d}{c})^{[n]}\}^{-1} \prod_{i=1}^{f} (\frac{d_i}{c})^{[n_i]}. \tag{A.5}$$

Poisson distribution

$k \in \mathbf{Z}_+,$
$x \geq 0,$

$$\pi_x(k) = \frac{1}{k!} x^k \exp(-x). \tag{A.6}$$

Negative binomial distribution

$k \in \mathbf{Z}_+,$
$0 < \alpha, 0 < \beta,$

$$n_{\alpha,\beta}(k) = \left(\begin{array}{c} \beta + k - 1 \\ k \end{array} \right) (\frac{1}{\alpha + 1})^\beta (\frac{\alpha}{\alpha + 1})^k. \tag{A.7}$$

Geometric distribution $n_{\overline{n},1}(k)$

$k \in \mathbf{Z}_+,$
$0 < \overline{n},$

$$G_{\overline{n}}(k) = \frac{1}{1 + \overline{n}} (\frac{\overline{n}}{1 + \overline{n}})^k. \tag{A.8}$$

β-distribution (conjugate to the binomial distribution and to the negative binomial distribution)

$p \in [0, 1],$
$0 < \alpha, 0 < \beta,$

$$b_{\alpha,\beta}(p) = \frac{\Gamma(\alpha + \beta)}{\Gamma(\alpha)\, \Gamma(\beta)} p^{\beta - 1} (1 - p)^{\alpha - 1}. \tag{A.9}$$

Dirichlet distribution (conjugate to the multinomial distribution)

$0 \leq p_i \leq i, 1 \leq i \leq f, \sum_{i=1}^f p_i = 1,$
$\underline{\alpha} = (\alpha_1, \ldots, \alpha_f) \in \mathbf{R}_+^f, 0 < \alpha_i,\ 1 \leq i \leq f, \sum_{i=1}^f \alpha_i = \alpha,$

$$\mathcal{D}_{\underline{\alpha}}(\mathbf{p}) = \frac{\Gamma(\alpha)}{\prod_{i=1}^f \Gamma(\alpha_i)} \prod_{i=1}^{f} p_i^{(\alpha_i - 1)}. \tag{A.10}$$

Γ-distribution (conjugate to the Poisson distribution)

$x \in \mathbf{R}_+$,
$0 < \alpha, 0 < \beta$,

$$\Gamma_{\alpha,\beta}(x) = \frac{1}{\Gamma(\beta)} \alpha^\beta \, x^{\beta-1} \exp(-\alpha \, x). \tag{A.11}$$

Exponential distribution $\Gamma_{\alpha,1}(x)$

$x \in \mathbf{R}_+$,
$0 < \alpha$,

$$\Gamma_{\alpha,1}(x) = \alpha \exp(-\alpha \, x). \tag{A.12}$$

A.2 Notation

General symbols.

Descending factorial moments, $x \in \mathbf{R}$ and $n \in \mathbf{Z}_+$,

$$x_{[n]} = x(x-1)\cdots(x-n+1), \quad x_{[0]} = 1. \tag{A.13}$$

Ascending factorial moments, $x \in \mathbf{R}$ and $n \in \mathbf{Z}_+$,

$$x^{[n]} = x(x+1)\cdots(x+n-1) = \Gamma(x+n)/\Gamma(x). \tag{A.14}$$

$[x]$	integer part of x
1_X	indicator function of the set X
δ_x	Dirac measure concentrated at x
$M_+^1(X)$	probability measures on X
$M_b^1(X)$	bounded, normalized measures on X
$C_b(X)$	bounded, continuous functions on X
$\mathrm{ex}(X)$	extreme points of the convex set X
$\mathrm{con}(X)$	convex hull of the set X
$\mathrm{cl}(X)$	closure of the set X
$\mathrm{lin}(X)$	linear span of the set X

Symbols for the description of the statistical scheme. The pages refer to the definition of the symbols.

Level–1

$\mathbf{J}$ configuration random variables, p.58.

$\mathbf{G}$ group configuration random variables, p.73.

Level–2

$\mathbf{K}$ occupation numbers, p.60.

$\mathbf{N}$ group occupation numbers, p.70, p.77.

Level–3

$\mathbf{Z}$ occupancy number random variables, p.64.

Limits I

$\mathcal{K}$ macroscopic occupation number random variable, p.82.

$\mathcal{M}$ macroscopic group occupation number random variable, p.82.

$\mathcal{E}$ energy density random variable, p.84.

$\mathcal{Z}$ macroscopic occupancy number random variable, p.85.

Limits II

$\mathbf{Q}$ particle density random variables, p.89, p.99.

$\mathcal{Q}$ macroscopic particle density random variable, p.103.

$\mathcal{N}$ macroscopic occupation number random variable, p.110.

$\underline{\mathcal{N}}$ limit of the renormalized macroscopic particle density, p.110.

Bibliography

1. L. Accardi and A. Bach. The harmonic oscillator as quantum central limit of Bernoulli processes. Technical report, University of Rome II, 1987.
2. L. Accardi and A. Bach. Quantum central limit theorems for squeezing operators. In L. Accardi and W. von Waldenfels, editors, *Lecture Notes in Mathematics 1396: Quantum Probability and Applications IV*, pages 7–19, Berlin, 1989. Springer.
3. L. Accardi and Y.G. Lu. A continuous version of de Finetti's theorem. *Ann. Prob.*, 21:1478–1493, 1993.
4. D.J. Aldous. Exchangeability and related topics. In P.L. Hennequin, editor, *Ecole d'Eté de Probabilités de Saint-Flour XIII – 1983*, pages 1–198, Berlin, 1985. Springer.
5. H. Araki and E.J. Woods. A classification of factors. *Pub. R.I.M.S. Kyoto Univ.*, 4:51–130, 1968.
6. F.T. Arecchi et al. Atomic coherent states in quantum optics. *Phys. Rev. A*, 6:2211–2237, 1972.
7. K.B. Athreya. On a characteristic property of Pólya's urn. *Stud. Sci. Math. Hung.*, 4:31–35, 1969.
8. A. Bach. On wave properties of identical particles. *Phys. Lett.*, 94A:251–254, 1984.
9. A. Bach. On the quantum properties of indistinguishable classical particles. *Lett. Nuovo Cimento*, 43:383–387, 1985.
10. A. Bach. On the interpretation of the fluctuations of black body radiation. *Phys. Lett. A*, 121:1–3, 1987.
11. A. Bach. Quanta and coherent states. *Lett. Math. Phys.*, 15:75–79, 1988.
12. A. Bach. Eine Fehlinterpretation mit Folgen: Albert Einstein und der Welle-Teilchen Dualismus. *Arch. Hist. Ex. Sci.*, 40:173–206, 1989.
13. A. Bach. Indistinguishable classical particles. Technical report, Haus Stapel, D-4409 Havixbeck, Germany, 1989.
14. A. Bach. On the statistics of nonclassical photon states. *Phys. Lett. A*, 134:405–408, 1989.
15. A. Bach. Boltzmann's probability distribution of 1877. *Arch. Hist. Ex. Sci.*, 41:1–40, 1990.
16. A. Bach. Indistinguishability, interchangeability and indeterminism. In D. Constantini and R. Cooke, editors, *Statistics in Science*, pages 149–166, Dordrecht, 1990. Kluwer.
17. A. Bach. De Finetti's theorem and Bell-type correlation inequalities. *Europhys. Lett.*, 16:513–518, 1991.
18. A. Bach. Why are independent bosons distributed according to Maxwell-Boltzmann statistics? *Europhys. Lett.*, 14:391–396, 1991.
19. A. Bach. Classification of indistinguishable particles. *Europhys. Lett.*, 21:515–520, 1993.

20. A. Bach. Nonclassical light is generated by finite systems of quanta. International Workshop on 'Quantum Communications and Measurement' (Nottingham), 1994.
21. A. Bach. Emergence of the simultaneous continuous and discrete structure of the electromagnetic field. *Rev. Math. Phys.*, 7:1–19, 1995.
22. A. Bach and U. Lüxmann-Ellinghaus. A simplex of probability measures associated with classical states of the harmonic oscillator. *Lett. Math. Phys.*, 9:103–106, 1985.
23. A. Bach and U. Lüxmann-Ellinghaus. The simplex structure of the classical states of the quantum harmonic oscillator. *Commun. Math. Phys.*, 107:553–560, 1986.
24. A. Bach and A. Srivastav. A characterization of the classical states of the quantum harmonic oscillator by means of de Finetti's theorem. *Commun. Math. Phys.*, 123:453–462, 1989.
25. P. Bártfai. On a conditional limit theorem. In J. Gani et al., editors, *Progress in Statistics, Volume I*, pages 85–91, Amsterdam, 1974. European Meeting of Statisticians, Budapest 1972, North-Holland.
26. P. Bártfai. Remarks on the exchangeable random variables. *Pub. Math. Debrecen*, 27:143–148, 1980.
27. H. Bauer. *Wahrscheinlichkeitstheorie und Grundzüge der Maßtheorie*. De Gruyter, Berlin, 3rd edition, 1978.
28. E. Beltrametti and G. Cassinelli. *The Logics of Quantum Mechanics*, volume 15 of *Encyclopedia of Mathematics and its Applications*. Addison–Wesley, Reading, 1981.
29. A. Benczur. On sequences of equivalent events and the compound Poisson process. *Stud. Sci. Math. Hung.*, 3:451–458, 1968.
30. D. Blackwell and D. Kendall. The Martin boundary for Pólya's urn scheme, and an application to stochastic population growth. *J. Appl. Prob.*, 1:284–296, 1964.
31. H. Blank. *Zur Statistik von ununterscheidbaren Teilchen*. PhD thesis, Westfälische–Wilhelms–Universität, Münster, 1985.
32. D.I. Blochinzew. *Grundlagen der Quantenmechanik*. Harri Deutsch, Frankfurt/M., 5th edition, 1966.
33. L. Boltzmann. Studien über das Gleichgewicht der lebendigen Kraft zwischen bewegten materiellen Punkten. *Sitz. Akad. Wiss. Wien, Math. Nat. Kl.*, 58:517–560, 1868. Reprinted in F. Hasenöhrl, editor, *Wissenschaftliche Abhandlungen*, Band I, pages 49-96, Barth, Leipzig, 1909.
34. L. Boltzmann. Über die Beziehung zwischen dem zweiten Hauptsatze der mechanischen Wärmetheorie, respective den Sätzen über das Wärmegleichgewicht. *Sitz. Akad. Wiss. Wien, Math. Nat. Kl.*, 76:373–435, 1877. Reprinted in F. Hasenöhrl, editor, *Wissenschaftliche Abhandlungen*, Band II, pages 164-223, Barth, Leipzig, 1909.
35. S.N. Bose. Plancks Gesetz und Lichtquantenhypothese. *Z. Physik*, 26:178–181, 1924.
36. W. Bothe. Zur Statistik der Hohlraumstrahlung. *Z. Physik*, 41:345–351, 1927.
37. L. Brillouin. Comparaison des différentes statistiques appliquées aux problèmes de quanta. *Ann. de Phys.*, 7:315–331, 1927.
38. L. Brillouin. *Die Quantenstatistik und ihre Anwendung auf die Elektronentheorie der Metalle*. Springer, Berlin, 1931.
39. G. Cassinelli and N. Zanghì. Conditional probabilities in quantum theory. I – conditioning with respect to a single event. *Nuovo Cimento*, 73B:237–245, 1983.

40. Y.S. Chow and H. Teicher. *Probability Theory.* Springer, New York, 2nd edition, 1988.
41. A.J. Coleman. Structure of fermion density matrices. *Rev. Mod. Phys.*, 35:668–687, 1963.
42. E.R. Davidson. *Reduced Density Matrices in Quantum Chemistry.* Academic Press, New York, 1976.
43. E.B. Davis. *Quantum Theory of Open Systems.* Academic Press, London, 1976.
44. B. de Finetti. Funzione caratteristica di un fenomeno aleatorio. *Atti R. Accad. Naz. Lincei, Cl. Fis. Mat. Nat., Memorie, Ser. 6*, 4:86–133, 1930.
45. B. de Finetti. La prévision: ses lois logiques, ses sources subjectives. *Ann. Inst. H. Poincaré*, 7:1–68, 1937.
46. B. de Finetti. Sulla proseguibilita di processi aleatori scambiabili. *Rendiconti dell'Istituto di Matematica dell'Università di Trieste*, 1:53–67, 1969.
47. B. de Finetti. *Teoria Della Probabilità.* Einaudi, Torino, 1970.
48. M. Delbrück. Was Bose–Einstein statistics arrived at by serendipity. *Journal of Chemical Education*, 57:467–470, 1980.
49. P. Diaconis. Finite forms of de Finetti's theorem on exchangeability. *Synthese*, 36:271–281, 1977.
50. P. Diaconis and D. Ylvisaker. Conjugate priors for exponential families. *Ann. Stat.*, 7:269–281, 1979.
51. J.M. Dickey. Conjugate families of distributions. In S.L. Kotz and N.L. Johnson, editors, *Encyclopedia of Statistics*, volume 2, pages 135–145. Wiley, N. York, 1982.
52. I.H. Dinwoodie and S.L. Zabell. Large deviations for exchangeable random vectors. *Ann. Prob.*, 20:1147–1166, 1992.
53. P.A.M. Dirac. On the theory of quantum mechanics. *Proc. Roy. Soc. A*, 112:611–677, 1926.
54. M. Dresden. The existence and significance of parastatistics. In K.W. Ford, editor, *Lectures on Astrophysics and Weak Interactions*, pages 377–469, New York, 1964. Brandeis Summer Institute in Theoretical Physics 1963, Gordon and Breach.
55. A. Einstein. Über einen die Erzeugung und Verwandlung des Lichtes betreffenden heuristischen Gesichtspunkt. *Ann. Phys.*, 17:132–148, 1905.
56. A. Einstein. Zum gegenwärtigen Stand des Strahlungsproblems. *Phys. Z.*, 10:185–193, 1909.
57. A. Einstein. Zur Quantentheorie der Strahlung. *Phys. Z.*, 18:121–128, 1917.
58. A. Einstein. Quantentheorie des einatomigen idealen Gases. *Sitz. Ber. Preuß. Akad. Wiss.*, pages 261–267, 1924.
59. A. Einstein. Quantentheorie des einatomigen idealen Gases, Zweite Abhandlung. *Sitz. Ber. Preuß. Akad. Wiss.*, pages 3–14, 1925.
60. R.S. Ellis. *Entropy, Large Deviations, and Statistical Mechanics.* Springer, New York, 1985.
61. G.G. Emch. *Algebraic Methods in Statistical Mechanics and Quantum Field Theory.* Wiley, New York, 1972.
62. W. Feller. *An Introduction to Probability Theory and Its Applications*, volume 1. Wiley, New York, 3rd edition, 1968.
63. W. Feller. *An Introduction to Probability Theory and Its Applications*, volume 2. Wiley, New York, 2nd edition, 1971.
64. E. Fermi. Sulla quantizzazione del gas perfetto monoatomico. *Rend. R. Accad. Lincei*, 3:145–149, 1926.
65. R.P. Feynman. *The Theory of Fundamental Processes.* Benjamin, New York, 1961.

66. W. Fleig. On the symmetry breaking mechanism of the strong-coupling BCS-model. *Acta Physica Austriaca*, 55:135–153, 1983.
67. O. Forster. *Analysis*, volume 3. Vieweg, Braunschweig, 1984.
68. S. Forte. Quantum mechanics and field theory with fractional spin and statistics. *Rev. Mod. Phys.*, 1992:193–236, 64.
69. D. Fürst. De Finetti: a scientist, a man. In G. Koch and F. Spizicchino, editors, *Exchangeability in Probability and Statistics*, pages 7–20, Amsterdam, 1982. North-Holland.
70. J. Galambos. Limit laws for mixtures with applications to asymptotic theory of extremes. *Z. Wahrscheinlichkeitstheorie verw. Gebiete*, 32:197–207, 1975.
71. J. Galambos. Exchangeable variates. In N.L. Johnson and S. Kotz, editors, *Urn Models and Their Application*, pages 97–106. Wiley, New York, 1977.
72. J. Galambos. *The Asymptotic Theory of Extreme Order Statistics*. Wiley, New York, 1978.
73. J. Galambos. Exchangeability. In S.L. Kotz and N.L. Johnson, editors, *Encyclopedia of Statistics*, volume 2, pages 573–577. Wiley, New York, 1982.
74. A. Galindo, A. Morales, and R. Nuñes–Lagos. Superselection principle and pure states of n–identical particles. *J. Math. Phys.*, 3:324–328, 1962.
75. A. Galindo and P. Pascual. *Quantum Mechanics*, volume 2. Springer, Berlin, 1991.
76. G. Gentile, j. Osservazioni sopra le statistiche intermedie. *Nuovo Cimento*, 17:493–497, 1940.
77. C.C. Gerry and J. Kiefer. Radial coherent states for the Coulomb problem. *Phys. Rev. A*, 37:665–671, 1988.
78. M.D. Girardeau. Permutation symmetry of many–particle wave functions. *Phys. Rev.*, 139:B500–B508, 1965.
79. R.J. Glauber. Optical coherence and photon statistics. In C. DeWitt, A. Blandin, and C. Cohen-Tannoudji, editors, *Quantum Optics and Electronics*, pages 64–185, Nw York, 1964. Gordon and Breach.
80. J. Glimm and A. Jaffe. *Quantum Mechanics*. Springer, Berlin, 1981.
81. A.V. Godambe. On representation of Poisson mixtures as Poisson sums and a characterization of the gamma distribution. *Math. Proc. Camb. Phil. Soc.*, 82:297–300, 1977.
82. A.B. Govorkov. Parastatistics and parafields. *Theor. Math. Phys.*, 54:234–241, 1983.
83. R.L. Graham, D.E. Knuth, and O. Patashnik. *Concrete Mathematics*. Addison–Wesely, Reading, second edition, 1989.
84. R. Haag. *Local Quantum Physics*. Springer, Berlin, 2ed edition, 1993.
85. J.B. Hartle, R.H. Stolt, and J.R. Taylor. Paraparticles of infinite order. *Phys. Rev.*, D2:1759–1760, 1970.
86. J.B. Hartle and J.R. Taylor. Quantum mechanics of paraparticles. *Phys. Rev.*, 178:2034–2051, 1969.
87. D. Heat and W. Sudderth. De Finetti's theorem on exchangeable variables. *Am. Stat.*, 30:188–189, 1976.
88. W. Heisenberg. Mehrkörperproblem und Resonanz in der Quantenmechanik. *Z. Physik*, 38:411–426, 1926.
89. W. Heitler and F. London. Wechselwirkung neutraler Atome und homöopolare Bindung nach der Qantenmechanik. *Z. Physik*, 44:455–472, 1927.
90. E. Hewitt and L.J. Savage. Symmetric measures on Cartesian products. *Trans. Amer. Math. Soc.*, 80:470–501, 1955.
91. B.M. Hill, D. Lane, and W. Sudderth. Exchangeable urn processes. *Ann. Prob.*, 15:1586–1592, 1987.

92. R.L. Hudson. Analogs of de Finetti's theorem and interpretative problems of quantum mechanics. *Found. Phys.*, 11:805–808, 1981.
93. R.L. Hudson and G.R. Moody. Locally normal symmetric states and an analogue of de Finetti's theorem. *Z. Wahrscheinlichkeitstheorie verw. Gebiete*, 33:343–351, 1976.
94. M. Jammer. *The Conceptual Development of Quantum Mechanics*. American Institute of Physics, New York, 2ed edition, 1989.
95. J.M. Jauch. *Foundations of Quantum Mechanics*. Addison-Wesley, Reading, 1968.
96. E.T. Jaynes. Some applications and extensions of the de Finetti representation theorem. In P. Goel and A. Zellner, editors, *Bayesian Inference and Decision Techniques*, pages 31–42. North-Holland, Amsterdam, 1986.
97. N.L. Johnson and S. Kotz. *Urn Models and Their Application*. Wiley, New York, 1977.
98. O. Kallenberg. A dynamical approach to exchangeability. In G. Koch and F. Spizzichino, editors, *Exchangeability in Probability and Statistics*, pages 87–96, Amsterdam, 1982. North-Holland.
99. D.G. Kendall. On finite and infinite sequences of exchangeable events. *Stud. Sci. Math. Hung.*, 2:319–327, 1967.
100. A. Khintchine. Sur les classes d'événements équivalents. *Math. Sb.*, 39:40–43, 1932.
101. J.F.C. Kingman. Uses of exchangeability. *Ann. Prob.*, 6:183–197, 1978.
102. J.R. Klauder and E.C.G. Sudarshan. *Fundamentals of Quantum Optics*. Benjamin, New York, 1968.
103. G. Koch and F. Spizzichino, editors. *Exchangeability in Probability and Statistics*. North-Holland, Amsterdam, 1982.
104. S. Kunte. The multinomial distribution, Dirichlet integrals and Bose–Einstein statistics. *Sankhyā*, 39A:305–308, 1977.
105. M.G.G. Laidlaw and C.M. de Witt. Feynman functional integrals for systems of indistinguishable particles. *Phys. Rev.*, D3:1375–1378, 1971.
106. L.D. Landau and E.M. Lifschitz. *Quantemechanik*. Akademie-Verlag, Berlin, 1967.
107. F.A. Lange. *Geschichte des Materialismus und seiner Bedeutung in der Gegenwart*, volume 2. Suhrkamp, Frankfurt, 1974. First edition 1866.
108. J.M. Leinaas and J. Myrheim. On the theory of identical particles. *Nuovo Cimento*, 37B:1–23, 1977.
109. M. Loève. *Probability Theory*. Van Nostrand, New York, 3rd edition, 1963.
110. H.A. Lorentz. Alte und neue Fragen der Physik. *Phys. Z.*, 11:1234–1257, 1910.
111. H.A. Lorentz. *Les théories statistiques en thermodynamique*. Teubner, Leipzig, 1916.
112. G. Lüders. Zum Symmetrisierungs–Postulat in der Quantenenmechanik identischer Teilchen. *Z. Physik*, 192:449–461, 1966.
113. E. Lukacs. *Characteristic Functions*. Griffin, London, 2nd edition, 1970.
114. Ja.P. Lumel'skiĭ. Random walks related to generalized urn schemes. *Soviet Math. Dokl.*, 14:628–632, 1973.
115. A. March. *Das neue Denken der modernen Physik*. Rowohlt, Hamburg, 1957.
116. A. Messiah. *Quantum Mechanics*, volume 2. North-Holland, Amsterdam, 1961.
117. A.M.L. Messiah and O.W. Greenberg. Symmetrization postulate and its experimental foundation. *Phys. Rev.*, 136:B248–B267, 1964.
118. P.-A. Meyer. *Quantum Probability for Probabilists*, volume 1538 of *Lecture Notes in Mathematics*. Springer, Berlin, 1993.

119. P.A. Meyer. Une remarque sur les lois echangeables. In J. Azéma, P.A. Meyer, and M. Yor, editors, *Séminaire de Probabilités XXIV 1988/89*, pages 486–487, Berlin, 1990. Springer. LNM 1426.
120. M.M. Miller and E.A. Mishkin. Representation of operators in quantum optics. *Phys. Rev.*, 164:1610–1617, 1967.
121. J.E. Moisman. The compound multinomial distribution, the multivariate β–distribution, and correlations among proportions. *Biometrika*, 49:65–82, 1962.
122. N.F. Mott. The exclusion principle and aperiodic systems. *Proc. Roy. Soc. London A*, 125:220–230, 1929.
123. W. Mückenheim. A review of extended probabilities. *Phys. Rep.*, 133:337–401, 1986.
124. L. Natanson. Über die statistische Theorie der Strahlung. *Phys. Z.*, 12:659–666, 1911.
125. K.R. Parthasarathy. *An Introduction to Quantum Stochastic Calculus*. Birkhäuser, Basel, 1992.
126. H. Paul. Interference between independent photons. *Rev. Mod. Phys.*, 58:209–231, 1986.
127. W. Pauli. Einsteins Beitrag zur Quantentheorie. In P.A. Schilpp, editor, *Einstein als Philosoph und Naturforscher*, pages 74–83. Vieweg, Braunschweig, 1979.
128. J.K. Percus. *Combinatorial Methods*. Springer, New York, 1971.
129. A. Perelomov. *Generalized Coherent States and Their Applications*. Springer, Berlin, 1986.
130. M. Planck. Ueber das Gesetz der Energieverteilung im Normalspectrum. *Ann. Phys.*, 4:553–563, 1901.
131. M. Planck. Die Gesetze der Wärmestrahlung und die Hypothese der elementaren Wirkungsquanten. In A. Eucken, editor, *Die Theorie der Strahlung und der Quanten. Verhandlungen auf einer von E. Solvay einberufenen Zusammenkunft (30. Oktober bis 3. November 1911)*, pages 77–94, Halle an der Saale, 1914. Wilhelm Knapp.
132. G. Pólya. Sur quelques points de la théorie des probabilités. *Ann. Inst. H. Poincaré*, 1:17–61, 1931.
133. J.M. Radcliffe. Some properties of coherent spin states. *J. Phys.*, A4:313–323, 1971.
134. M. Reed and B. Simon. *Methods of Modern Mathematical Physics*, volume 1. Academic Press, New York, 1972.
135. B. Saleh. *Photoelectron Statistics*. Springer, Berlin, 1978.
136. L.I. Schiff. *Quantum Mechanics*. McGraw-Hill, New York, 3rd edition, 1968.
137. O. Steinmann. Symmetrization postulate and cluster property. *Nuovo Cimento*, 44A:755–767, 1966.
138. R.H. Stolt and J.R. Taylor. Classification of paraparticles. *Phys. Rev.*, D1:2226–2228, 1970.
139. E. Størmer. Symmetric states on infinite tensor products of C^*-algebras. *J. Functional Analysis*, 3:48–68, 1969.
140. R.F. Streater and A.S. Wightman. *PCT, Spin & Statistics And All That*. Benjamin, New York, 1964.
141. E.C.G. Sudarshan and N. Mukunda. *Classical Mechanics: A Modern Perspective*. Wiley, New York, 1974.
142. S. Sýkora. Quantum theory and the Bayesian inference problems. *J. Stat. Phys.*, 11:17–27, 1974.
143. R.L. Taylor, P.Z. Daffer, and R.F. Patterson. *Limit Theorems for Sums of Exchangeable Random Variables*. Rowman & Allanheld, Totowa, 1985.

144. D. ter Haar. *Elements of Thermostatics*. Holt, Rinehart and Winston, New York, 1966.
145. I. Vincze. On the maximum probability principle in statistical physics. In J. Gani et al., editors, *Progress in Statistics, Volume II*, pages 869–893, Amsterdam, 1974. European Meeting of Statisticians, Budapest 1972, North-Holland.
146. I. Vincze. Some problems in connection with the Bose-Einstein statistic. *Sankhyā*, 37B:355–362, 1975.
147. I. Vincze. Indistinguishability of particles and exchangeable random variables. In *Proceedings of the seventh Conference on Probability Theory (Brasov, 1982)*, pages 393–398, Utrecht, 1985. VNU Sci. Press.
148. M. von Laue. Die Einsteinschen Energieschwankungen. *Verh. d. Deutsch. Phys. Ges.*, pages 198–202, 1915.
149. R. von Mises. *Wahrscheinlichkeitsrechnung*. Deuticke, Leipzig, 1931.
150. R. von Mises. Über Aufteilungs- und Besetzungswahrscheinlichkeiten. *Rev. Fac. Sci. Univ. Istanbul*, 4(5):145–163, 1939. Reprinted in Ph. Frank et. al., editors, *Selected Papers of Richard von Mises*, Volume 2, pages 313–334, American Mathematical Society, Providence, 1964.
151. J. von Plato. De Finetti's earliest works on the foundation of probability. *Erkenntnis*, 31:263–282, 1989.
152. S. Weinberg. Unified theories of elementary-particle interaction. *Scientific American*, 213:50–59, 1974.
153. H. Weyl. *Quantenmechanik und Gruppentheorie*. Hirzel, Leipzig, 2ed edition, 1931.
154. E.P. Wigner. Über nicht kombinierende Terme in der neueren Quantentheorie. Zweiter Teil. *Z. Physik*, 40:883–892, 1927.
155. H.P. Yuen. Two-photon coherent states of the radiation field. *Phys. Rev. A*, 13:2226–2243, 1976.
156. S.L. Zabell. W. E. Johnson's 'sufficientness' postulate. *Ann. Stat.*, 10:1091–1099, 1982.

Druck: STRAUSS OFFSETDRUCK, MÖRLENBACH
Verarbeitung: SCHÄFFER, GRÜNSTADT

Lecture Notes in Physics

For information about Vols. 1–444
please contact your bookseller or Springer-Verlag

Vol. 445: J. Brey, J. Marro, J. M. Rubí, M. San Miguel (Eds.), 25 Years of Non-Equilibrium Statistical Mechanics. Proceedings, 1994. XVII, 387 pages. 1995.

Vol. 446: V. Rivasseau (Ed.), Constructive Physics. Results in Field Theory, Statistical Mechanics and Condensed Matter Physics. Proceedings, 1994. X, 337 pages. 1995.

Vol. 447: G. Aktaş, C. Saçlıoğlu, M. Serdaroğlu (Eds.), Strings and Symmetries. Proceedings, 1994. XIV, 389 pages. 1995.

Vol. 448: P. L. Garrido, J. Marro (Eds.), Third Granada Lectures in Computational Physics. Proceedings, 1994. XIV, 346 pages. 1995.

Vol. 449: J. Buckmaster, T. Takeno (Eds.), Modeling in Combustion Science. Proceedings, 1994. X, 369 pages. 1995.

Vol. 450: M. F. Shlesinger, G. M. Zaslavsky, U. Frisch (Eds.), Lévy Flights and Related Topics in Physics. Proceedigs, 1994. XIV, 347 pages. 1995.

Vol. 451: P. Krée, W. Wedig (Eds.), Probabilistic Methods in Applied Physics. IX, 393 pages. 1995.

Vol. 452: A. M. Bernstein, B. R. Holstein (Eds.), Chiral Dynamics: Theory and Experiment. Proceedings, 1994. VIII, 351 pages. 1995.

Vol. 453: S. M. Deshpande, S. S. Desai, R. Narasimha (Eds.), Fourteenth International Conference on Numerical Methods in Fluid Dynamics. Proceedings, 1994. XIII, 589 pages. 1995.

Vol. 454: J. Greiner, H. W. Duerbeck, R. E. Gershberg (Eds.), Flares and Flashes, Germany 1994. XXII, 477 pages. 1995.

Vol. 455: F. Occhionero (Ed.), Birth of the Universe and Fundamental Physics. Proceedings, 1994. XV, 387 pages. 1995.

Vol. 456: H. B. Geyer (Ed.), Field Theory, Topology and Condensed Matter Physics. Proceedings, 1994. XII, 206 pages. 1995.

Vol. 457: P. Garbaczewski, M. Wolf, A. Weron (Eds.), Chaos – The Interplay Between Stochastic and Deterministic Behaviour. Proceedings, 1995. XII, 573 pages. 1995.

Vol. 458: I. W. Roxburgh, J.-L. Masnou (Eds.), Physical Processes in Astrophysics. Proceedings, 1993. XII, 249 pages. 1995.

Vol. 459: G. Winnewisser, G. C. Pelz (Eds.), The Physics and Chemistry of Interstellar Molecular Clouds. Proceedings, 1993. XV, 393 pages. 1995.

Vol. 460: S. Cotsakis, G. W. Gibbons (Eds.), Global Structure and Evolution in General Relativity. Proceedings, 1994. IX, 173 pages. 1996.

Vol. 461: R. López-Peña, R. Capovilla, R. García-Pelayo, H. Waelbroeck, F. Zertuche (Eds.), Complex Systems and Binary Networks. Lectures, México 1995. X, 223 pages. 1995.

Vol. 462: M. Meneguzzi, A. Pouquet, P.-L. Sulem (Eds.), Small-Scale Structures in Three-Dimensional Hydrodynamic and Magnetohydrodynamic Turbulence. Proceedings, 1995. IX, 421 pages. 1995.

Vol. 463: H. Hippelein, K. Meisenheimer, H.-J. Röser (Eds.), Galaxies in the Young Universe. Proceedings, 1994. XV, 314 pages. 1995.

Vol. 464: L. Ratke, H. U. Walter, B. Feuerbach (Eds.), Materials and Fluids Under Low Gravity. Proceedings, 1994. XVIII, 424 pages, 1996.

Vol. 465: S. Beckwith, J. Staude, A. Quetz, A. Natta (Eds.), Disks and Outflows Around Young Stars. Proceedings, 1994. XII, 361 pages, 1996.

Vol. 466: H. Ebert, G. Schütz (Eds.), Spin – Orbit-Influenced Spectroscopies of Magnetic Solids. Proceedings, 1995. VII, 287 pages, 1996.

Vol. 467: A. Steinchen (Ed.), Dynamics of Multiphase Flows Across Interfaces. 1994/1995. XII, 267 pages. 1996.

Vol. 468: C. Chiuderi, G. Einaudi (Eds.), Plasma Astrophysics. 1994. VII, 326 pages. 1996.

Vol. 469: H. Grosse, L. Pittner (Eds.), Low-Dimensional Models in Statistical Physics and Quantum Field Theory. Proceedings, 1995. XVII, 339 pages. 1996.

Vol. 470: E. Martínez-González, J. L. Sanz (Eds.), The Universe at High-z, Large-Scale Structure and the Cosmic Microwave Background. Proceedings, 1995. VIII, 254 pages. 1996.

Vol. 471: W. Kundt (Ed.), Jets from Stars and Galactic Nuclei. Proceedings, 1995. X, 290 pages. 1996.

Vol. 472: J. Greiner (Ed.), Supersoft X-Ray Sources. Proceedings, 1996. XIII, 350 pages. 1996.

Vol. 473: P. Weingartner, G. Schurz (Eds.), Law and Prediction in the Light of Chaos Research. X, 291 pages. 1996.

Vol. 474: Aa. Sandqvist, P. O. Lindblad (Eds.), Barred Galaxies and Circumnuclear Activity. Proceedings of the Nobel Symposium 98, 1995. XI, 306 pages. 1996.

Vol. 475: J. Klamut, B. W. Veal, B. M. Dabrowski, P. W. Klamut, M. Kazimierski (Eds.), Recent Developments in High Temperature Superconductivity. Proceedings, 1995. XIII, 362 pages. 1996.

Vol. 476: J. Parisi, S. C. Müller, W. Zimmermann (Eds.), Nonlinear Physics of Complex Systems. Current Status and Future Trends. XIII, 388 pages. 1996.

Vol. 477: Z. Petru, J. Przystawa, K. Rapcewicz (Eds.), From Quantum Mechanics to Technology. Proceedings, 1996. IX, 379 pages. 1996.

New Series m: Monographs

Vol. m 1: H. Hora, Plasmas at High Temperature and Density. VIII, 442 pages. 1991.

Vol. m 2: P. Busch, P. J. Lahti, P. Mittelstaedt, The Quantum Theory of Measurement. XIII, 165 pages. 1991. Second Revised Edition: XIII, 181 pages. 1996.

Vol. m 3: A. Heck, J. M. Perdang (Eds.), Applying Fractals in Astronomy. IX, 210 pages. 1991.

Vol. m 4: R. K. Zeytounian, Mécanique des fluides fondamentale. XV, 615 pages, 1991.

Vol. m 5: R. K. Zeytounian, Meteorological Fluid Dynamics. XI, 346 pages. 1991.

Vol. m 6: N. M. J. Woodhouse, Special Relativity. VIII, 86 pages. 1992.

Vol. m 7: G. Morandi, The Role of Topology in Classical and Quantum Physics. XIII, 239 pages. 1992.

Vol. m 8: D. Funaro, Polynomial Approximation of Differential Equations. X, 305 pages. 1992.

Vol. m 9: M. Namiki, Stochastic Quantization. X, 217 pages. 1992.

Vol. m 10: J. Hoppe, Lectures on Integrable Systems. VII, 111 pages. 1992.

Vol. m 11: A. D. Yaghjian, Relativistic Dynamics of a Charged Sphere. XII, 115 pages. 1992.

Vol. m 12: G. Esposito, Quantum Gravity, Quantum Cosmology and Lorentzian Geometries. Second Corrected and Enlarged Edition. XVIII, 349 pages. 1994.

Vol. m 13: M. Klein, A. Knauf, Classical Planar Scattering by Coulombic Potentials. V, 142 pages. 1992.

Vol. m 14: A. Lerda, Anyons. XI, 138 pages. 1992.

Vol. m 15: N. Peters, B. Rogg (Eds.), Reduced Kinetic Mechanisms for Applications in Combustion Systems. X, 360 pages. 1993.

Vol. m 16: P. Christe, M. Henkel, Introduction to Conformal Invariance and Its Applications to Critical Phenomena. XV, 260 pages. 1993.

Vol. m 17: M. Schoen, Computer Simulation of Condensed Phases in Complex Geometries. X, 136 pages. 1993.

Vol. m 18: H. Carmichael, An Open Systems Approach to Quantum Optics. X, 179 pages. 1993.

Vol. m 19: S. D. Bogan, M. K. Hinders, Interface Effects in Elastic Wave Scattering. XII, 182 pages. 1994.

Vol. m 20: E. Abdalla, M. C. B. Abdalla, D. Dalmazi, A. Zadra, 2D-Gravity in Non-Critical Strings. IX, 319 pages. 1994.

Vol. m 21: G. P. Berman, E. N. Bulgakov, D. D. Holm, Crossover-Time in Quantum Boson and Spin Systems. XI, 268 pages. 1994.

Vol. m 22: M.-O. Hongler, Chaotic and Stochastic Behaviour in Automatic Production Lines. V, 85 pages. 1994.

Vol. m 23: V. S. Viswanath, G. Müller, The Recursion Method. X, 259 pages. 1994.

Vol. m 24: A. Ern, V. Giovangigli, Multicomponent Transport Algorithms. XIV, 427 pages. 1994.

Vol. m 25: A. V. Bogdanov, G. V. Dubrovskiy, M. P. Krutikov, D. V. Kulginov, V. M. Strelchenya, Interaction of Gases with Surfaces. XIV, 132 pages. 1995.

Vol. m 26: M. Dineykhan, G. V. Efimov, G. Ganbold, S. N. Nedelko, Oscillator Representation in Quantum Physics. IX, 279 pages. 1995.

Vol. m 27: J. T. Ottesen, Infinite Dimensional Groups and Algebras in Quantum Physics. IX, 218 pages. 1995.

Vol. m 28: O. Piguet, S. P. Sorella, Algebraic Renormalization. IX, 134 pages. 1995.

Vol. m 29: C. Bendjaballah, Introduction to Photon Communication. VII, 193 pages. 1995.

Vol. m 30: A. J. Greer, W. J. Kossler, Low Magnetic Fields in Anisotropic Superconductors. VII, 161 pages. 1995.

Vol. m 31: P. Busch, M. Grabowski, P. J. Lahti, Operational Quantum Physics. XI, 230 pages. 1995.

Vol. m 32: L. de Broglie, Diverses questions de mécanique et de thermodynamique classiques et relativistes. XII, 198 pages. 1995.

Vol. m 33: R. Alkofer, H. Reinhardt, Chiral Quark Dynamics. VIII, 115 pages. 1995.

Vol. m 34: R. Jost, Das Märchen vom Elfenbeinernen Turm. VIII, 286 pages. 1995.

Vol. m 35: E. Elizalde, Ten Physical Applications of Spectral Zeta Functions. XIV, 228 pages. 1995.

Vol. m 36: G. Dunne, Self-Dual Chern-Simons Theories. X, 217 pages. 1995.

Vol. m 37: S. Childress, A.D. Gilbert, Stretch, Twist, Fold: The Fast Dynamo. XI, 410 pages. 1995.

Vol. m 38: J. González, M. A. Martín-Delgado, G. Sierra, A. H. Vozmediano, Quantum Electron Liquids and High-T_c Superconductivity. X, 299 pages. 1995.

Vol. m 39: L. Pittner, Algebraic Foundations of Non-Commutative Differential Geometry and Quantum Groups. XII, 469 pages. 1996.

Vol. m 40: H.-J. Borchers, Translation Group and Particle Representations in Quantum Field Theory. VII, 131 pages. 1996.

Vol. m 41: B. K. Chakrabarti, A. Dutta, P. Sen, Quantum Ising Phases and Transitions in Transverse Ising Models. X, 204 pages. 1996.

Vol. m 42: P. Bouwknegt, J. McCarthy, K. Pilch, The $\mathcal{W}_3$ Algebra. Modules, Semi-infinite Cohomology and BV Algebras. XI, 204 pages. 1996.

Vol. m 43:

Vol. m 44: A. Bach, Indistinguishable Classical Particles. VIII, 157 pages. 1997.

Vol. m 45: M. Ferrari, V. T. Granik, A. Imam, J. C. Nadeau (Eds.), Advances in Doublet Mechanics. XVI, 214 pages. 1997.